"This extraordinarily lucid book is essentially a massive experiment with basic philosophic conceptions. Accordingly, it operates on many levels, and offers varied and rich facets of interest. Yet it is unified by impressive directness of intent and style, and by an intricate and able skill in concocting a remarkable array of historical and factual materials with a complex dialectic of ideas."

—The Personalist

"Although the concept of metaphor is of considerable philosophical importance, surprisingly little good work has been done on the subject. It is therefore an important and welcome publishing event when a work as scholarly, well written, and provocative as Professor Turbayne's apears."

—The Philosophical Review

". . . a clear and brilliant answer to the age-old question, how can we best account for the relationship that seems to exist between mirror-image and object."

—Dalhousie Review

"Turbayne's book is an extremely well written, lucid study of the role of metaphor in the attempted solutions of scientific and philosophical problems. At the outset, however, it is important to understand that by 'metaphor' he intends rather more than we would in the normal use of the word."

—International Philosophical Quarterly

THE MYTH OF METAPHOR

The Myth of Metaphor

REVISED EDITION

by COLIN MURRAY TURBAYNE
with Forewords by
MORSE PECKHAM and FOSTER TAIT
and an Appendix by
ROLF EBERLE

University of South Carolina Press
Columbia, South Carolina

For my mother

ALICE EVA RENE TURBAYNE

and in memory of my father

DAVID LIVINGSTONE TURBAYNE

Preface to the Revised Edition

A revised edition of a book often contains revisions in the text that reflect revisions in the thought of the author. Just as often, however, when this occurs, the character of the book is changed. A writer, like a painter or a sculptor, can spoil his work by tampering with it. Perhaps because he can never recapture that same mood of the agony and ecstasy of composition he can transform what was a unity into a disunity. Frequently during the years since my book was published I have been astonished by the revelation of the various ideas that many readers have found in it and of the different interpretations that they have given to it. They have seen things that I had not seen.

For these reasons I have judged it better to leave the text alone except for making minor corrections and to let others say what they find significant in the original.

In his Foreword, Morse Peckham, with the same flair that he displays in his *Man's Rage for Chaos* and his recent *Art and Pornography,* stresses the book's significance for conceptions of poetry in the theory and practice of literary criticism. In a second Foreword, Foster Tait happily accentuates the relevance of my conception of metaphor to metaphysics. In the Appendix, Rolf Eberle, with enviable precision and insight, explicates my conceptions of model and metaphor in terms of formal model-theory and clarifies my attempt to illustrate the use of an extended metaphor as a tool for scientific discovery.

As an author I am enormously heartened to have my book revised and renewed in this way.

I am deeply grateful to Morse Peckham and Brand Blanshard for their generous support and encouragement.

C. M. T.

Villino Buonamici
Firenze
March 1970

Contents

PART THREE

TESTING THE METAPHOR

Foreword I

I think that I shall never forget my pleasure and excitement when I first read Professor Turbayne's *The Myth of Metaphor*. I found it delicious merely as a piece of writing; its elegance, lucidity, wit, and charm make every page of a very tough-minded philosophical discourse into a continuously flowing gratification. But of course it was what the book says that makes it as significant to the student of literature as to the student of philosophy. As it happened, I was particularly interested when I saw it listed in a catalogue of recent scholarly publications because I myself had just published an essay on metaphor. "A Little Plain Speaking on a Weary Subject" was its subtitle, a phrase that indicated plainly enough how dissatisfied I was with all accounts of metaphor I had come across. I was, to be sure, after bigger game than merely metaphor. I was stalking a common conception of poetry which I felt to be totally in error.

That doctrine of poetics is, to use a term innovated by Professor E. D. Hirsch, Jr., "the doctrine of semantic autonomy." It is a doctrine so unacceptable, indeed so silly, that I scarcely know any longer how to state it. Yet it is probably the notion on poetic language now most widely held by professional critics and literary scholars. It is the notion that poetry is a unique mode of discourse. As I put it in my own essay on metaphor, "The most common position today, the position that carries with it the richest critical and academic status and self-approval, is that poetry is a means of discovering a 'truth' which is accessible to no other way of thinking, and that the technique of such thinking is metaphor. Ultimately, it follows, and at the profoundest level, poetry is metaphor." Metaphor = poetry = truth was the equation I was out to hunt down and destroy, if I could. For I was convinced that metaphor is not only a normal semantic mode but a mode essential for the existence and above all the extension of the semantic functions of language. It is the only way we have for saying something new.

In sum, I was aware that metaphor was not a literary problem but a philosophical one. Hence my delight in discovering that Professor Turbayne had dealt with it philosophically, and with a richness and precision that, as a literary student, I could not hope to equal. As a general rule, philosophers are not much help to literary critics, for literature presents semantic problems of a complexity far beyond anything philosophers try to deal with. That is one reason we students of literature flounder about so desperately. Here, however, was a philosopher who was dealing directly with a problem that not only is of the greatest importance to literary critics, but of the highest significance in disposing of a notion of poetry which for some decades had done intolerable harm to the theory and practice of literary criticism. Professor Turbayne opens the way for a lucid and rational discussion of literature, and to the discovery of its *real* problems, some of which we are now beginning to be aware of.

The *Myth of Metaphor* is a book I try to persuade all of my graduate students to read. And now that it is back in print I shall do my best to get them to buy it.

Morse Peckham
Department of English Language and Literature
University of South Carolina

Foreword II

The professional critic or literary scholar who equates metaphor with poetry and poetry with truth is both disesteemed and
opposed by the philosopher who, after deciding that metaphors
are literally false, or cognitively insignificant, dismisses them as
"mere" instances of semantic confusion. But both critic and philosopher perform a disservice. For the critic is as guilty of over-
stating the semantic function of metaphor as the philosopher is of
understating it. In short, critic and philosopher alike have failed
to appreciate either the importance or the limitations of metaphor.
The author of *The Myth of Metaphor* is guilty of neither of these
infractions. His approach to the subject is both sympathetic and
tough. It is surely one of the most original, critical, suggestive, and
thoroughly rewarding approaches that I have encountered.

My initial experience with Professor Turbayne's book came at a
time when I was attempting to complete my requirements for the
doctorate. I was busy, and since my own areas of primary philosophical interest lay outside aesthetics, I experienced some mis-
givings about taking time from my dissertation to read the book,
even though it had been recommended to me by a fellow student
whom I respected quite highly. What finally persuaded me to read
The Myth of Metaphor was the knowledge that its author had stud-
ied with the same professor who had succeeded in making me realize that an analysis of metaphor could be of considerable value to
people working in epistemology and the philosophy of science.
Unfortunately, however, the majority of essays on metaphor that
I had read appeared to contain nothing of genuine interest for the
epistemologist or philosopher of science. Imagine my delight
when I discovered that Professor Turbayne had succeeded in
treating the subject in a manner which was of at least as much
importance and interest to the professional philosopher as it was
to the scholar of literature.

Detecting metaphors and understanding how they are distinguished from literal uses of language is one thing, and it is important; but knowing how to use metaphors without being used by them is another thing, and it may be even more important. A major portion, and in my opinion the most exciting and original portion of *The Myth of Metaphor* is devoted to a clarification of how one may avoid being victimized by metaphor. At this point Professor Turbayne's concern about being "captured" by metaphor strikes me as similar to Descartes' fear of being victimized by custom when in *Meditation I* he says:

> But it is not sufficient to have made these remarks, we must also be careful to keep them in mind. For these ancient and commonly held opinions still revert frequently to my mind, long and familiar custom having given them the right to occupy my mind against my inclination and rendered them almost masters of my belief.

A metaphor which is used for purposes of illustration or explanation may become a model; and when it does, there are definite disadvantages to be gained from taking it literally. For example, to take as literally true one of the numerous models which occur in science would be to attribute the model to the world. It would be to say, in effect, that the world "really is that way." But this is not a useful way to approach either the world, knowledge, or science. It is not because it minimizes the possibility of employing other models to illustrate the facts in question. Professor Turbayne attempts to prevent this species of "utilization by metaphor" by demonstrating how models may be substituted for other models. For his purpose he selects the geometrical model of the universe used by Descartes, among others, and opposes it with a carefully developed linguistic model. The metaphor being attacked makes the stakes high because the metaphor is actually the basis for an influential metaphysics. It is, in fact, closely related to causal interpretations of nature.

I would be doing an injustice to Professor Turbayne's remarkable book if I failed to mention its contribution to the literature devoted to the writings of George Berkeley. In his work entitled *An Essay towards a New Theory of Vision*, Berkeley again and again compares vision to a language. Near the end of this work the author actually identifies vision with a "Universal Language of Nature." Such repeated emphasis on language, in both metaphor and simile, indicates that Berkeley set considerable store by his model, perhaps too much. Turbayne's extensive development of a language model clarifies and even improves Berkeley's own choice of a model, a model which Berkeley invented and placed in competition with the traditional geometrical model of vision. In fact, Professor Turbayne's model may even provide a "bridge" between Berkeley's works on vision and his *A Treatise Concerning the Principles of Human Knowledge*. The significance for his metaphysics of these works on vision, and particularly of the *Essay*, has long been a debated issue among interpreters of Berkeley's philosophy.

Foster Tait
Department of Philosophy
University of South Carolina

THE MYTH OF METAPHOR

Introduction

In ORDER to illustrate the facts, to control them more effectively, to induce attitudes, or to inculcate ways of behavior, artists, philosophers, theologians, and scientists have used various devices. An extraordinarily successful one often used to illuminate areas that might otherwise have remained obscure is the model or metaphor. Its use involves the pretense that something is the case when it is not. Hobbes pretended that the state was a many-jointed monster or leviathan; Shakespeare that it was a hive of honey bees, "Creatures that by a rule in nature teach the act of order to a peopled kingdom." Plato, however, presented the obscure facts of human nature as if they were luminous facts about the state. Descartes pretended that the mind in its body was the pilot of a ship; Locke that it was a room, empty at birth but full of furniture later; and Hume that it was a theatre. Theologians have pretended that the relation between God and man is that of father to son. Optical theorists have pretended that we see by geometry. Metal experts present the facts about metals that break after constant use as if they suffer fatigue, while physicists make believe at some times that light moves in waves, at others that it consists of corpuscles, in order to account for different observable facts in the motion of light.

Just as often, however, the pretense has been dropped, either by the pretenders or by their followers. There is a difference between using a metaphor and taking it literally, between using a model and mistaking it for the thing modeled. The one is to make believe that something is the case; the other is to believe that it is. The one is to use a disguise or mask for illustrative or explanatory purposes; the other is to mistake the mask for the face. Both the pretense and the mistake involve, in the words of Gilbert Ryle, "the presentation of the facts belonging to one category in the idioms appropriate to another." Both thus involve the crossing of different sorts. But while the former is to represent the facts of one sort *as if* they belong to another, the latter is to claim that they actually belong. While the former adds nothing obviously to the actual process, the latter involves the addition of features that are the products of speculation or invention instead of discovery. It thus involves the insinuation of metaphysics. The juxtaposition of the titles "metaphor" and "metaphysics" in the pages of the *Encyclopaedia Britannica* has a significance richer than that intended by the editors.

The history of science may be treated from the point of view that it records attempts to place metaphysical disguises upon the faces of process and procedure. After the disguise or mask has been worn for a considerable time it tends to blend with the face, and it becomes extremely difficult to "see through" it. We can still penetrate the obvious masks like the Pilot, the Theatre, the Wave, and the Corpuscle. But in others the make-up is hidden. Ryle saw through the Ghost in the Machine, and may have laid it. Freud penetrated the Father-image. Some have lifted the disguises only to replace them with fresh ones. Of these some have been aware of what they were doing; others have been taken in by their own devices. Deluded to think that they were ejecting metaphysics, they were actually replacing metaphysical theories that they found disagreeable by others more agreeable to them.

Nevertheless, the crossings of different sorts often have great illustrative or explanatory value. It is not necessarily a confusion to present items belonging to one sort in the idioms appropriate to another. If it were, we should have to say that the making of every myth, of every new metaphor, and of almost every theory involved a confusion. On the other hand, it is a confusion to present the items of one sort in the idioms of another—without awareness. For to do this is not just to cross two different sorts; it is to confuse them. It is to mistake, for example, the theory for the fact, the procedure for the process, the myth for history, the model for the thing, and the metaphor for the face of literal truth. Accordingly, to expose a categorial confusion, to explode a myth, or to "undress" a hidden metaphor is not just to re-allocate the items: it is to show that these sometimes valuable fusions are actually confusions. This is a large part of what I try to show in the following pages.

With this purpose, I try to explode the metaphysics of mechanism. This I do, *first*, by exposing mechanism as a case of being victimized by metaphor. Descartes and Newton I choose as excellent examples of metaphysicians of mechanism *malgré eux*, that is to say, as unconscious victims of the metaphor of the great machine. These two great "sort-crossers" of our modern epoch have so imposed their arbitrary allocation of the facts upon us that it has now entered the coenesthesis of the entire Western World. Together they have founded a church, more powerful than that founded by Peter and Paul, whose dogmas are now so entrenched that anyone who tries to re-allocate the facts is guilty of more than heresy; he is opposing scientific truth. For the accepted allocation is now identified with science. All this is so in spite of the meager opposition offered by the theologians, a few poets, and fewer philosophers, who, in general, have been victimized by their own metaphors to the same degree as their rivals. They have opposed one metaphysics, done without awareness, by another. They have been operating on the wrong level.

Secondly, I try to show that the metaphysics of mechanism can be dispensed with. The best way to do this is to show that it is only a metaphor; and the best way to show this is to invent a new metaphor. I therefore treat the events in nature *as if* they compose a language, in the belief that the world may be treated just as well, if not better, by making believe that it is a universal language instead of a giant clockwork. But my purpose in presenting the language metaphor is not so much to build as to destroy, not so much to plant flowers as to pull up weeds. I present the language metaphor not because I want to make a new myth in the hope that, although the generation to whom it is first told cannot possibly believe it, the next may, and the generations after. I present it in order to begin to show: first, that there is a remedy against the domination imposed, not by generals, statesmen, and men of action, whose power dissolves when they retire, but by the great sort-crossers, whose power increases when they die —the remedy provided by becoming aware of metaphor; and second, that the metaphysics that still dominates science and enthralls the minds of men is nothing but a metaphor, and a limited one. With this purpose in turn, I proceed to test these two metaphors. Having concluded that the competition can be conducted between different metaphors, I find that tests rarely resorted to reveal themselves. But it would be premature to apply them without first putting the metaphors to work within a specific subject matter where they can be used to produce different theories capable of being tested by those canons customarily used to test any scientific theory. Accordingly, I test the father of the machine model, namely, the geometrical model, against the language model in the concrete problem of vision, a subject still so dogmatically illustrated—as it has been for more than two thousand years—by the geometrical model that the mask is mistaken for the face. I try to show that the language model peculiarly illuminates this ancient problem of how we see, shedding a bright light on dark areas dimly lit by its great rival.

In order to develop my theme I needed two sorts of illustration, the one showing what it is to use metaphor, the other what it is to be used by it. Victims of metaphor were easy to find, for they have been legion. Although I might have exhibited more recent case histories, I chose those of Descartes and Newton for the reasons already given. It was more difficult to find illustrations of the sustained use of metaphor, for there are hardly any to be had. The chief candidates, so far as I was concerned, were Freud's *Interpretation of Dreams*, Plato's *Republic*, and Berkeley's two essays on vision. These masterpieces offer the nearest approximations known to me of the deliberate and sustained application of an extended metaphor—that is, a model—to a concrete problem. But they are only approximations. Their authors eventually fell into the same trap as all the other victims. Having invented their new metaphors, and having used them with great skill, apparently with awareness of what they were doing, they were then so beguiled by the charm of their creations that they mistook these interpretations for the things interpreted. They took their own metaphors literally, their make-believe for the real thing. It was as though general contentment had been given a uniform.

Nevertheless, any one of these could have been remade to satisfy one of my purposes as a counter-illustration of the abuse of an extended metaphor. But only one might be made to satisfy all my purposes. I wanted a metaphor rich in connotation, powerful and flexible in application, and familiar to all, that would have a chance of competing with the machine in the particular and in the general: by showing up its weak illumination of a particular problem and by suggesting an alternative myth. It seemed to me that the language metaphor might fill these requirements. Accordingly, I chose Berkeley's attempt to represent the facts of one sort (vision) in the idioms appropriate to another (language). This he made in two short works: *An Essay towards a New Theory of Vision*, in 1709, and *The Theory of Vision Vindicated and Explained*, in 1733.[1]

In Chapters IV, V, and VIII of my book, I adopt much of this theory as well as much of Berkeley's account of the properties he specified in his model, while making modifications to suit my needs. In order to avoid tediousness in presentation, I give references to Berkeley's writings only at the ends of paragraphs.[2] My main modifications are as follows: While he kept the language metaphor well hidden, I bring it into the open. Moreover, while he fell into inconsistencies—especially a grave one at the heart of the theory, concerning the visible and the tangible square, which spoiled his theory—I try to remove them. Finally, he fell into the inevitable trap. Having pretended that vision is a language, and having kept up this pretense with great success, he dropped it: "Vision is the language of the Author of Nature." It was as though, finding the old sorts of theological argument outmoded, he decided to offer a modern parable which, in the telling, ceased to be a parable and became literal truth. What I try to do is to keep the metaphor vividly alive throughout.

C. M. T.

University of Rochester
March 1961

1. See Colin Murray Turbayne, ed., Berkeley: *Works on Vision*, New York: Bobbs-Merrill, 1963.
2. Abbreviations are as follows: A *Alciphron;* E *An Essay towards a New Theory of Vision;* H *Three Dialogues between Hylas and Philonous;* I Introduction and Draft Introduction to the *Principles;* P *The Principles of Human Knowledge;* PC *Philosophical Commentaries;* S *Siris;* V *The Theory of Vision.*

PART ONE

FINDING THE METAPHOR

CHAPTER I

The Nature of Metaphor

1. Using Metaphor

HOWEVER appropriate in one sense a good metaphor may be, in another sense there is something inappropriate about it. This inappropriateness results from the use of a sign in a sense different from the usual, which use I shall call "sort-crossing." Such sort-crossing is the first defining feature of metaphor and, according to Aristotle, its genus:

> Metaphor (*meta-phora*) consists in giving the thing a name that belongs to something else; the transference *(epi-phora)* being either from genus to species, or from species to genus, or from species to species, *or* on the grounds of analogy.[1]

Thus metaphor is logically indistinguishable from trope, the use of a word or phrase in a sense other than that which is proper to it. Aristotle apparently regarded the features that distinguish metaphor from trope as psychological, but he did not specify them. What these differentiae are, which make the chief problem of metaphor, I shall consider shortly.

Notice how wide Aristotle's definition is. Metaphor comprehends all those figures that some distinguish as: synechdoche (sort-crossing from genus to species or *vice versa*, for example, treating the university as of the same sort as a building on the campus, or the division as a battalion or battery); metonymy (giving the thing a name that belongs to an attribute or adjunct, for example, "sceptre" for authority, or "the crown" for the sovereign); catachresis (giving the thing which lacks a proper name a

---

1. *Poetics,* 1457, my italics.

name that belongs to something else, which includes the use of ordinary words in a technical sense, for example, Berkeley's "idea" and Whitehead's "point"); and metaphor (giving the thing which already has a proper name a name that belongs to something else on the grounds of analogy, for example, "That utensil!" applied to Mussolini by Churchill, "metal fatigue," and "Man is a wolf"). Aristotle's discernment in treating analogical sort-crossing, to which many restrict the name "metaphor," as only one species of metaphor, is commendable. That the resemblance theory of metaphor is inadequate has been cogently argued by Max Black who concludes that it would be more illuminating to say that in some cases "the metaphor creates the similarity than to say that it formulates some similarity antecedently existing." [2] Metaphor often is grounded on such similarity but need not be. It might be grounded on revelation.

Wide as Aristotle's definition is I make it wider. Without stretching his meaning unduly, I interpret his singular "name" to mean either a proper name, a common name, or a description expressible as a phrase, a sentence, or even a book. In which case a more adequate presentation of the defining feature of metaphor I am now considering is made by Gilbert Ryle. Metaphor consists in "the presentation of the facts of one category in the idioms appropriate to another." [3] As with Aristotle's definition the fundamental notion expressed here is that of transference from one sort to another or, for short, of sort-crossing. Thus defined, metaphor comprehends some cases of the model and also of the allegory, the parable, and the enigma which, from Aristotle's point of view, would be cases of extended metaphor. I should point out, however, that although this definition admirably suits my purposes, it is not Ryle's definition of metaphor at all. It is, indeed, his definition of category-mistake, and is therefore basic to his correction of the dominant modern theory of mind.

The definition is still not wide enough. Some cases of metaphor may not be expressed in words. Again without stretching Aris-

2. Max Black, "Metaphor," *Aristotelian Society Proceedings* (1954–55), 284–85.
3. Gilbert Ryle, *The Concept of Mind* (London: Hutchinson's University Library, 1949), p. 8.

totle's meaning unduly, I interpret his "name" to mean a sign or a collection of signs. This will allow artists who "speak" in paint or clay to "speak" in metaphor. Michelangelo, for example, used the figure of Leda with the swan to illustrate being lost in the rapture of physical passion, and the same figure of Leda, only this time without the swan, to illustrate being lost in the agony of dying. It will also allow the concrete physical models of applied scientists, the blackboard diagrams of teachers, the toy blocks of children that may be used to represent the battle of Trafalgar, and the raised eyebrow of the actor that may illustrate the whole situation in the state of Denmark, to be classified as metaphor.

Such an extraordinary extension of the meaning may seem absurd. This, the traditional view, is expressed by W. Bedell Stanford: "If the term *metaphor* be let apply to every trope of language, to every result of association of ideas and analogical reasoning, to architecture, music, painting, religion, and to all the synthetic processes of art, science, and philosophy, then indeed metaphor will be warred against by metaphor . . . and how then can its meaning stand?" [4] This objection to the identification of metaphor with trope is, of course, valid. It therefore seems to be valid against Aristotle's definition and my extension of it which, because they blur needed distinctions, should be discarded. But all that follows is that sort-crossing, by itself, is not enough to define metaphor. Other features are needed. Aristotle probably realized this. Apparently noticing no logical distinction between metaphor and trope, he correctly gave the trope as the genus of metaphor, and then correctly distinguished its various species which were different logical kinds of sort-crossing. But this is where he stopped. Having offered one defining feature, he was desiderating others that constitute a new dimension. Not every sort-crossing is a metaphor, but every sort-crossing is potentially a metaphor. We need this initial width which I now proceed to whittle down.

The use of metaphor involves the pretense that something is the case when it is not. That pretense is involved is only sometimes disclosed by the author. Descartes said: "I have hitherto described this earth, and generally the whole visible world, *as if it*

---

4. W. Bedell Stanford, *Greek Metaphor* (Oxford: Basil Blackwell, 1936), p. 103.

were merely a machine." [5] But just as often metaphors come un-labeled. Michelangelo made no meta-sculptural inscription on the base of the sculpture "Night" disclosing that he had borrowed the Leda figure to illustrate the nature of death, although the initiated knew that he had. Nor did Churchill say that he was using the name "utensil" figuratively.

That pretense is involved is not revealed by the grammar. There is no significant difference grammatically between each of these pairs:

The timber-wolf is a wolf	Man is a wolf
That frying-pan is a utensil	Mussolini is a utensil
Ordinary language	Visual language
Muscle fatigue	Metal fatigue

Yet I am inclined to say that only the second members involve metaphor. In the examples the two different things or sorts referred to in each pair share the same name, and it is as if they share other things as well. When I say that the timber-wolf is a wolf I am actually giving to timber-wolves a name that belongs to other wolves, and I mean that the timber-wolf is a sort included in the larger sort *wolf;* or that timber-wolves share with other wolves all the defining properties of "wolf"; or that timber-wolves are included in the denotation of "wolf." On the other hand, when I say that man is a wolf (metaphorically speaking) I am actually giving him a name shared by all other wolves just as if I believe that he is another sort of wolf like the timber-wolf or the Tasmanian wolf, sharing with them all the defining properties of "wolf," or sharing with them inclusion in the denotation of "wolf." But though I give him the same name I do not believe he is another sort of wolf. I only make believe he is. My words are not to be taken literally but only metaphorically. That is, I pretend that something is the case when it is not, and I implicitly ask my audience to do the same.

But more clarity is needed on the matter of what is pretense and what is not. Certainly when I use a metaphor there is no pretense about the name-transference. Man actually shares the name

5. *Principles,* IV. 188.

"wolf." But it is pretense that man is a sort of wolf. However, something besides the name is transferred from wolves to men. I do not merely *pre*tend that man shares the properties of wolves; I *in*tend it. What these properties are I may, but need not, specify. They cannot be all the properties common to wolves, otherwise I should intend that man is actually a wolf. Thus when I say that man is a wolf (metaphorically speaking) I intend that he shares some of the properties of wolves but not enough of them to be classified as an actual wolf—not enough to let him be ranged alongside the timber-wolf and the Tasmanian wolf. Or when I say that vision is a language I intend that vision shares some of the properties of language but not enough to let it be ranged alongside English and French.

Consider, however, the effect of conjoining the metaphors listed above with their corresponding literal expressions:

> Men and timber-wolves are wolves
> Mussolini and that frying-pan are utensils
> Visual and ordinary language
> Metal and muscle fatigue

These conjunctions serve to disclose the ambiguity of certain words. When I speak of the timber- and the Tasmanian wolf I refer to two different sorts of wolves by using the word "wolf" in only one sense. But when I speak of the timber-wolf and the man-wolf it is not a case of referring to two different sorts of wolves. This is so because the word "wolf" is being used in two different senses even though it is as it were being used in only one. The two different senses are called the literal and the metaphorical senses of "wolf." Similar remarks apply to the other examples. It seems then that conjunctions such as these are devices for making us aware of the duality of meaning of one word or sign. Can it be then that awareness of the duality of meaning is the test of using metaphor? But consider some other similar conjunctions:

> Seeing the point of the needle and the joke
> Smelling of musk and insolence
> Clad only in her tiara and an embarrassed expression
> A toast to general contentment and General de Gaulle

Such conjunctions reveal the mechanics of the traditional joke that depends upon using the same word in two different senses at once. A common way of treating them is to force the "literal" meaning: to say, for example, that to smell of musk and "My Sin" is to smell of two different sorts of perfume, while to smell of insolence is not to be smelly at all. Moreover, just like the earlier given set, these conjunctions reveal not only the presence of duality of sense but the presence of metaphor, one sense being literal, the other metaphorical. In which case such expressions as "seeing the point of a joke" and "smelling of insolence" are metaphors. But this is to confuse the literal with the physical and to hold that any non-physical sense is metaphorical. Originally the literal may have been nothing but the physical. But it is not so now. There may be many literal senses, only some of which are physical. This seems to be the case here. While we are aware of the many senses in which words like "see," "general," etc., may be used, we do not usually consider that one of these senses—the physical— holds a monopoly of the literal sense, the others being figurative. Accordingly, it is more accurate to say that such conjunctions reveal the presence of duality of sense or sort-crossing but not necessarily the presence of metaphor.

It is harder to construct conjunctions producing the same effect with words in which the original—probably physical—sense has been lost to all but classicists:

> Comprehend the meaning
> Perceive the table
> Discourse upon logic

For the same reason it is harder to make metaphors with them. Nevertheless it is possible for the expressions just listed and for those like "see the point of the joke" etc., treated above, to become metaphorical. Awareness of duality of meaning, as we have seen, is not enough to do it. Neither my awareness of my ability to see the point of a needle at the same time as I say that I can see the point of a joke, nor my awareness of my ability to detect the smell of musk at the same time as I say that he came ino the room smelling of insolence, is enough to make my assertions metaphorical. Nor, if I am an etymologist, is my awareness of the gross original sense of "comprehend" at the same time as I say that I comprehend your meaning. In all these cases I am using words in one or other of their literal senses. I am representing the facts of one sort in words that may be equally appropriate to the facts of another. What more then is needed to make these expressions metaphorical?

The answer lies in the *as if* or *make-believe* feature already sketched and illustrated. When Descartes says that the world is a machine or when *I* say with Seneca that man is a wolf, and neither of us intends our assertions to be taken literally but only metaphorically, both of us are aware, *first,* that we are sort-crossing, that is, re-presenting the facts of one sort in the idioms appropriate to another, or, in other words, of the duality of sense. I say *"are* aware," but of course, we *must* be, otherwise there can be no metaphor. We are aware, *secondly,* that we are treating the world and man as if they belong to new sorts. We are aware of the duality of sense in "machine" and "wolf," but we make believe that each has only one sense—that there is no difference in kind, only in degree, between the giant clockwork of nature and the pygmy clockwork of my wrist watch, or between man-wolves and timber-wolves. It is as if the sentences

> The world and this clock are machines

and

> Men and timber-wolves are wolves

which we both know to be absurd, were meaningful and true. In short, the use of metaphor involves both the awareness of duality of sense and the pretense that the two different senses are one. To Dr. Johnson's remark that metaphor "gives you two ideas for one," we need to add that it gives you two ideas *as* one.

To encompass this new factor Aristotle's definition must therefore be expanded. Gilbert Ryle offers a still better definition of metaphor: "It represents the facts . . . *as if* they belonged to one logical type or category (or range of types of categories), when they actually belong to another." [6] This greatly illuminates the subject of metaphor because it draws our attention to those

6. *The Concept of Mind,* p. 16, my italics.

two features that I have been stressing, namely sort-crossing or the fusion of different sorts, and the pretense or *as if* feature. I should again point out, however, that although this definition is about the best definition of metaphor known to me, it is not Ryle's definition of metaphor at all. It is, indeed, his alternative definition of category-mistake or categorial confusion.

The definition of metaphor now arrived at is based on Aristotle's. It improves his, first, by widening it to include as potential metaphors any signs possessing duality of meaning; and secondly, by narrowing it to differentiate metaphor from trope, thus satisfying the desideratum in Aristotle's definition. Moreover, it saves his definition, which is basically correct, first, by retaining his notion that metaphor is the genus of which all the rest are species; and, secondly, by preserving the distinctions between metonymy, synecdoche, and catachresis. But, of course, none of these is properly a metaphor until the *as if* prescription is filled. All of them have the basic ingredient of sort-crossing or duality of sense, and, to this extent, are all tropes. All of them, however, are potentially metaphors. Any trope can achieve full metaphorhood but only for that user who fuses the two senses by making believe there is only one sense. Thus to the plain man there may be no metaphor in Aristotle's "substance," Descartes' "machine of nature," Newtonian "force" and "attraction," Thomas Young's "kinetic energy," and Michelangelo's figure of Leda. Placed in their customary contexts these present to him nothing but the face of literal truth. To the initiated, however, who are aware of the "gross original" senses as well as the now literal senses, they may become metaphors. There are no metaphors *per se*.

Few differentiae need be added to provide adequate definitions of the model, the parable, the fable, the allegory, and the myth. What is special about these which, according to Aristotle, would be special cases of metaphor? I said earlier that when we use a metaphor it is left open whether we specify what properties are to be transferred from one sort to another. Max Black clarifies this feature.[7] To call a man a "wolf" metaphorically without specifying the wolf-properties implies that the "literal uses of the word normally commit the speaker to acceptance of a set of

7. "Metaphor," 286, 289.

standard beliefs about wolves," for example, that they are fierce, hungry, scavengous, and so on. These "current platitudes" he calls "the system of associated commonplaces." On the other hand, the properties may be specified: "Metaphors can be supported by specially constructed systems of implications, as well as by accepted commonplaces; they can be made to measure and need not be reach-me-downs." Now I should say that when the latter occurs, we are dealing with that special instance of metaphor called the "model." For example, one may use the language model to illustrate the subject of vision by drawing on such *accepted commonplaces* as signs, things signified, rules of grammar, etc., and by *specifying* that the connection between signs and things signified is first arbitrary, then conventional, that the only signs to be used are nouns and exclamations, and so on. Or, for example, one may make a map of a particular terrain that omits marks for towns, roads, and railways. Or, to choose another example, Plato devoted more than half of his so-called *Republic* to the construction of his intricate system of implications. The details specified in his model of the state are so complex and unusual that posterity, mistaking the model for the thing modeled, misnamed the whole work.

The fable, the parable, the allegory, and the myth are, like the model, extended or sustained metaphors. The wave model of light, Aesop's fables, Jesus' parable of the sower, Bunyan's *Pilgrim's Progress,* the Napoleonic myth, and the myth of volitions have this much in common: All of them as Jean de la Fontaine said of the fable, "are not what they appear"; they are all cases of representing the facts that belong to one sort as if they belonged to another; they are stories that we make believe to be true. After this there are differences. The myth—that most peculiar case of metaphor belonging properly to a later stage in its life—I shall discuss in more detail later. As for the others, like the model, which need not be grounded on resemblance to the thing modeled—for it may create it—they can satisfy the phrase "on the grounds of analogy" in Aristotle's definition but need not. Like the model, they are supported by specially constructed systems of implication. Aesop not only chose his animals with care but also selected with forethought their requisite characteristics. Moreover, as with the use of models, the audience is sometimes told, though often only implicitly, sometimes not, that metaphor is involved. Finally, like the model, they are offered and meant to be understood in "the spirit of this serious make-believe [in which] not only the little girl talks about her dolls but we ourselves read our Dante." [8] This is so even though the literal meaning may not be serious. Great clowning may be merely funny to the child, but to the adult who gets both meanings it is serious funniness. "And so amused," said Jean de la Fontaine of the fable, "we learn."

Unlike the model, however, the fable, the parable, and the allegory are all designed to inculcate better behavior by creating right attitudes or dispositions, to all of which the literal meaning of the story told is merely incidental. Unlike the myth, which grows like a tree, all are deliberate inventions. The vehicle for them all is usually a fictitious narrative which we make believe is true, though the allegory and parable may be expressed in argumentative discourse, and the allegory in drama, paint, clay, or mime. It is, however, the story told together with the lesson delivered, contributing first the duality of sense and then the duality plus unity, that make all these cases of metaphor and sometimes great works of art. In the case of the fable, the fabulist explicitly discloses the presence of metaphor in a statement of higher type that may come suddenly at the end. In the others it is usually left to the audience to provide its own interpretation. In the myth it is left to posterity.

While the fable is the appropriate adjunct or replacement of moral discourses—"we yawn at sermons"—the parable and the allegory may appropriately adjoin or replace theo-moral discourses. And while the fable uses animals, flowers, birds, and bees which we make believe can speak, act, and feel like persons, the imagery of the parable is more conservative. Animals may be used but they retain their animal identity. The allegory, not so restricted, may even use abstract ideas like goodness and charity which we personify. Shakespeare's "So work the honey bees" is a fable pointing a different moral from Mandeville's, while "Let us drink a toast to General Contentment" may start an allegory if we put the general on horseback or give him a uniform.

8. J. A. Stewart, *The Myths of Plato* (London: Macmillan, 1905), p. 7.

2. Being Used by Metaphor

I now consider some of the effects of a good metaphor. It is this aspect that prompted Aristotle to say, after praising Euripides at the expense of Aeschylus: "The greatest thing by far is to ⸱e a master of metaphor. It is the one thing that cannot be learned from others. It is the mark of genius." [9] The great sort-crossers from Pythagoras through Plato, Descartes, and Newton to Einstein have changed our attitudes to the facts. How have they done this? Consider first some of the metaphors used to illustrate metaphor.

Kenneth Burke says a metaphor offers a "perspective": "Metaphor is a device for seeing something in terms of something else. . . . A metaphor tells us something about one character considered from the point of view of another character. And to consider A from the point of view of B is, of course, to use B as a *perspective* upon A." [10] Bedell Stanford, who stresses the two-ideas-as-one aspect, describes metaphor as "the stereoscope of ideas," because it achieves "this *integration* of diversities." [11] Black's meta-metaphors, however, are more illuminating. They accommodate the feature of attitude-shift, and are, therefore, harbingers of things to come in this book. An effective metaphor, he says, acts as a *screen* through which we look at the world; or it *filters* the facts, suppressing some and emphasizing others. It "brings forward aspects that might not be seen at all through another medium." The chess metaphor, for example, used to illustrate war, emphasizes the game-of-skill features while it suppresses the grimmer ones. A good metaphor produces thereby "shifts in attitudes." [12]

The attitude-shifts produced by an effective metaphor point to a later stage in its life. A story often told—like advertising and propaganda—comes to be believed more seriously. Those details stressed tend to stay stressed while those suppressed tend to stay suppressed until another effective metaphor restores them. What was once a new occasional sort produced by the original sort-cross-

---

9. *Poetics*, 1459.
10. Kenneth Burke, *A Grammar of Motives* (New York: Prentice-Hall, Inc., 1945), pp. 503–04.
11. *Greek Metaphor*, pp. 101, 105.
12. "Metaphor," 287–88.

ing merges into a conventional sort. A change in attitudes to the facts can even issue in a change in fact. When the facts are re-allocated, and this re-allocation becomes acceptable to many, the old allocations are neglected, and the facts are changed. A rose by some other name might not smell as sweet. The tomato re-allocated to the fruit class changes its taste history. A Dry-Martini health drink loses its flavor. To neglect the conventional allocation: "Locke-Berkeley-Hume," in our history courses, and to adopt the re-allocation: "Newton-Berkeley-Mach," is to alter first, our attitude to Berkeley, then Berkeley. The human characteristics that Aesop pretended were owned by animals have become literally part of their equipment. We no longer make believe that foxes are cunning and lambs gentle. They are. It will need another Aesop to make bulls cruel and lions murderous. When the pretense is dropped either by the original pretenders or their followers, what was before called a *screen* or *filter* is now more appropriately called a *disguise* or *mask*. There is a difference between using a metaphor and being used by it, between using a model and mistaking the model for the thing modeled. The one is to make believe that something is the case; the other is to believe it.

In order to sharpen the distinction I have begun to make, let us return to our definition of metaphor: the presentation of the facts of one sort as if they belonged to another. This is Ryle's definition of category-mistake. But it seems altogether unlikely that Ryle regards metaphors as mistakes, for such "category-mistakes" may have great value. It is not necessarily a mistake to cross sorts. If it were, we should have to say that the making of every metaphor and model, as well as every explanation of something in terms of something else, involved a mistake. On the other hand, it is a mistake to present the facts of one sort in the idioms of another without awareness. For to do this is not just to fuse two different senses of a sign; it is to confuse them. Thus my distinction may be characterized as that between a category-fusion and a category-confusion or, to borrow another word from Ryle, between sort-crossing and sort-"trespassing." Accordingly, being used by a metaphor or taking a metaphor literally is a case of sort-trespassing.

Just as I began with sort-crossing to find out what it is to use metaphor, so I begin with sort-trespassing to find out what it is to be used by metaphor. Not all cases of sort-crossing involve metaphor, and not all cases of sort-trespassing involve taking a metaphor literally. But in each case the reverse is true. There need be no metaphor in the use of such utterances as: "I see the point of the joke," and "Let us drink a toast to general contentment." Although some will say that "to see the table" shows the literal sense of "see" while "to see the point of the joke" shows the metaphorical sense, in my view these expressions merely reveal different literal senses of "see," one having to do with the use of eyes, the other not, while the latter may become metaphorical. Those who use such expressions are usually aware that there are other senses of "see" and "general." There is, therefore, sort-crossing without metaphor. Correspondingly, the use of such expressions need not involve sort-trespassing. But if the user is puzzled over the sharpness of the point of the joke or wonders whether the general is a corps commander or only a divisional commander, then we know at once that he has trespassed. For he has confused two different but customary literal senses of these words. He has confused them because he has been unaware of the duality of sense involved. For him, the proposal: "Let us drink a toast to General de Gaulle and General Contentment" is by no means absurd. It is a proposal to do honor to two different generals. Only complex computing machines will escape sort-trespassing.

When does sort-trespassing become a case of taking a metaphor literally? Only when one of the two different senses confused is metaphorical and this is taken for the literal. He who says "Man is a wolf," metaphorically speaking, is aware of the duality of senses and merely makes believe that man is a wolf. But he who is taken in by the metaphor is unaware and believes man is a wolf. For him the class of wolves is enlarged by the addition of another subclass. For him it is not a case of different senses of the word "wolf"; it is a case merely of different sorts of wolves. In which case for him the assertion that timber-wolves and men are wolves contains no absurdity. Thus taking a metaphor literally is a special instance of sort-trespassing. But since a metaphor is not a metaphor *per se* but only for someone, from one point of view it is better to say that sometimes the metaphor is not noticed; it is hidden. That is, if X is aware of the metaphor while Y is not, X says that Y is being taken in by the metaphor, or being used by it, or taking it literally. But for Y it is not a case of taking the metaphor literally at all, because for him there is no metaphor. He is speaking literally or taking it literally. Similarly in the case of models, X says that Y takes the model for the thing, while for Y there is no model. The model is the thing.

There are three main stages in the life of a metaphor. At first a word's use is simply inappropriate. This is because it gives the thing a name that belongs to something else. It is a case of misusing wòrds, of "going against ordinary language," and, therefore, of breaking the conventions. All the great sort-crossers were unconventional. Great metaphors are no better or no worse off in this regard than the ordinary mistakes in naming. When children, scientists, and philosophers call camels "dogs," sounds "vibrations," numbers "classes of similar classes," aggression "inferiority," meanings "ways of testing," mind "behavior," and tables "ideas," they all do something inappropriate because unconventional. The child who mistakenly calls a camel "a dog" gets as near to or as far from literal truth as the mechanist who calls the human body "a machine." This is precisely the case with every effective metaphor. Our first response is to deny the metaphor and affirm the literal truth: "Bodies attract each other" (only people attract or are attracted, though bodies may move together) ; "Look around the world—you will find it to be nothing but one great machine" (only clocks, sewing machines, and automobiles are machines, though the world moves and has an intricate arrangement of parts); "Metal fatigue and the cruel sea" (only people suffer fatigue and are cruel, though metals wear out after constant use, and harmless people may drown in a rough sea).

But because such affirmation and denial produce the required duality of meaning, the effective metaphor quickly enters the second stage in its life; the once inappropriate name becomes a metaphor. It has its moment of triumph. We accept the metaphor by acquiescing in the make-believe. This is the stage at which— by making believe that camels are dogs, sounds are vibrations, mind is behavior, the human body is a machine, forces reside in bodies, bodies attract each other, the Russian boundary is an iron curtain, and so on—the metaphor is used by us with awareness to illuminate obscure or previously hidden facts. This is the stage at which the metaphor, being new, fools hardly anyone.

The moments of inappropriateness and triumph are short compared to the infinitely long period when the metaphor is accepted as commonplace. The last two stages are sometimes described as the transition from a "live" metaphor to one "moribund" or "dead." But it is better to say that either the metaphor is now hidden or it ceases to be one. Within this long period the original metaphor may develop in various ways only one of which is a case of taking metaphor literally.

To speak of "comprehending" meanings and hailing a "cab" is to speak appropriately and usually literally. It is not necessarily to speak in metaphor or to take a metaphor literally. In these cases the original sense has been lost to most of us. How then can we take a metaphor literally (which involves the confusion of different senses) if there is only one sense? Nevertheless an etymologist, who associates physical seizure with the first and the characteristic smell of the male goat with the second, may relish the metaphors. If a philosopher thinks that the mind *acts* to *grasp* what it only *comprehends*, that is, if he thinks that the act of comprehending a universal or abstract idea is a special sort of action and that grasping with the hands is another sort, then he is engaged in sort-trespassing, while to the etymologist just mentioned, he takes a metaphor literally.

Similarly, to speak of "smelling" of insolence or of "seeing" the point of a joke is to speak appropriately and usually literally. It is not necessarily to speak in metaphor or to reveal that one is captured by metaphor. In these cases, though the original sense has not been lost, we remain aware that there are two senses, and do not confuse them. Nevertheless, we may treat them as metaphors. If, like children, who, as Plato said, "cannot judge what is allegorical and what is literal," we are puzzled about what sort of perfume insolence resembles, then again we trespass and, from the point of view of him who sees the metaphor, we are indeed captured by it.

In this part of the third stage of metaphor we no longer make believe that camels are dogs, that sounds are vibrations, etc. Camels are now nothing but dogs; sounds are nothing but vibrations, and the human body is nothing but a machine. What had before been models are now taken for the things modeled. That is, special sets of implications had been invented for dogs, vibrations, and machines, designed to explain the facts about camels, sounds, and human bodies in dog, vibration, and machine language. Conclusions about one were reducible to the premises about the other. But now reducibility has become reductionism, for camels and dachshunds are now literally different sorts of dogs. The machine metaphor has become mechanism, for human bodies and clockwork now differ only in degree, not in kind.

What accounts for this transition from using metaphor with awareness to being used or victimized by it, from make-believe to belief? I do not pretend to offer a complete explanation, but it seems likely that it lies, in large part at least, in the Principle of Association, on the full-scale use of which Hume based his philosophical prestige. The explanation of reductionism or the Reductive Fallacy by this principle is still acceptable. If so, then the same principle should also accommodate the phenomenon of taking a metaphor literally. For the latter is a special case of sort-trespassing or reductionism. The long continued association of two ideas, especially if the association has theoretical and practical benefit, tends to result in our confusing them. Students find no difficulty in treating sound as nothing but vibrations. In the case of metaphor, the confusion is aided by the following factors: First, the two ideas already share the same name, a factor of great power in producing the belief in identity. Second, we are not always told that the two ideas are really different. If Newton was using "attraction" as a metaphor, he did not say so. Third, even when we are told, we export properties from one idea to the other, such being the nature of metaphor. Finally, the line between make-believe and belief is thin. Prospero's wicked brother, at first merely play-acting, came to "credit his own lie." He did believe he was indeed the Duke of Milan. These factors supplement the major factor of constant association or use which almost alone can change the stamp of nature.

Whether I describe these two attitudes as using metaphor and being used by it, or as awareness of metaphor and lack of awareness of it; or whether the metaphor involved is to be called myth, model, fable, or allegory; it is certain that the two attitudes exist. The history of science and philosophy records many instances of both, but many more of the latter. One is the attitude of the Wizard of Oz himself for whom "The Emerald City" is but a name and the green glasses but a colored screen to heighten the make-believe. The other is the attitude of his duped subjects for whom the Emerald City is really green. Forgetful of their green glasses, they believe they contemplate nothing but the un-made-up face of the truth. While, on the one hand, the use of metaphor to illuminate dark areas—the price of which use is constant vigilance —is not to confuse a device of procedure with elements of the process, on the other hand, being used by metaphor involves the addition to the process of features of procedure that are the products of invention or speculation. The victim of metaphor accepts one way of sorting or bundling or allocating the facts as the only way to sort, bundle, or allocate them. The victim not only has a special view of the world but regards it as the only view, or rather, he confuses a special view of the world with the world. He is thus, unknowingly, a metaphysician. He has mistaken the mask for the face. Such a victim who is a metaphysician *malgré lui* is to be distinguished from that other metaphysician who is aware that his allocation of the facts is arbitrary and might have been otherwise.

CHAPTER II

Analysis and Synthesis

1. Foreword

IN THIS CHAPTER I try to do two things. *First,* in order to give content to my distinction between using metaphor and being used by it, I offer some illustrations of actual victims of metaphor from the history of science. My main examples are Descartes and Newton, two scientists who first invented or developed procedures for describing the process of nature and then confused ingredients of their procedures with the process they described. *Second,* in order to find a method that may help me to avoid the errors of these giants, I present their methods in some detail.

These two are peculiarly appropriate for my purposes. More than others in modern times they have influenced the attitudes of subsequent scientists, philosophers, and ordinary people, so that our vision of the world remains largely a complication of the Newtonian and the Cartesian. Moreover, they were both philosophers of science as well as scientists, having left us not only their own descriptions of nature but also higher level accounts of what they thought they were doing in making these descriptions. Finally, in spite of their mistaken beliefs about what they thought they were doing, their methods worked.

In different degrees Descartes and Newton were aware and unaware of what they were doing. To some degree they did not confuse the ingredients of their procedures with the process they described. But they also thought that many ingredients of procedure were duplicated in the process. They thus added qualities to the world, thinking that these ingredients were not just inventions or decisions in the realm of procedure but actual discoveries of fact.

In this respect they were like cooks who first use a recipe with great skill and then add the pages of their recipe to the stew.

To change the picture, they were in part victims of Bacon's Idols of the Theatre, "because in my judgment all the received systems are but so many stage-plays representing worlds of their own creation . . . neither only of entire systems but also of many principles and axioms in science which by tradition, credulity, and negligence have come to be received." [1] I shall show how these scientists confused their "stage-plays" with the events they depicted in the way Bacon describes. Then, following the lead given by Descartes, I shall try to do away with the "stage-plays": "The sciences now have masks on them; if the masks were taken off they would appear supremely beautiful." [2] Finally I shall show how the "stage-plays" may be re-performed, or the "masks" put back, but with this crucial difference, with awareness that they are only "stage-plays" or only "masks."

The best way to understand the methods of these scientists is to see them as inherited from a most ancient tradition known for about two thousand years as the double procedure of analysis and synthesis. And the best way to understand this procedure is to start with the account given by its great progenitor, Plato. Indeed we shall see that the modifications made upon the original by those who came later are slight.

2. Analysis and Synthesis

The scientific procedures now in vogue were invented by the Greeks. Since their time scientists have been trying to rediscover them, to improve upon them, and to apply them. In order, first, to discover truth or to solve concrete problems in practical matters and, secondly, to present their discoveries or solutions, the Greeks devised two distinct procedures. To discover truth they invented inductive argument considered as the means whereby general conclusions or principles could be derived from the facts. To present their discoveries they invented the axiomatic method in which from axioms and definitions they derived theorems by deduction.

1. *Novum Organum* (London, 1626), 1.44.
2. *Private Thoughts* (Composed 1619).

These two procedures were subsequently called by a variety of
Greek and Latin names, the most common being the Greek
"analysis" and "synthesis" and the Latin "resolution" and "composition." Although the names "analysis" and "synthesis" were
commonly used in these senses until the nineteenth century, there
is now some risk of confusion in using them in their old senses.
This is because Kant unhappily decided to use the names "analytic" and "synthetic" both in the old way and in a way diametrically opposed to the old. As a result, it is now fashionable to call
deduction "analytic" and induction "synthetic." Whatever names
are chosen, the two procedures have been used with different
degrees of rigor by scientists since Plato's time. Thus Descartes'
Meditations, Newton's *Opticks,* Berkeley's *Essay on Vision,* and
Kant's *Prolegomena* were presented in the inductive or analytical
manner, while Plato's *Republic,* Euclid's *Elements* and *Catoptrics,*
Spinoza's *Ethics,* Descartes' *Principles* (in part), Newton's *Principia,* Berkeley's *Vindication,* and Kant's three *Critiques* were
presented in the deductive or synthetical style. Of these, Euclid's
application of the deductive method to geometry and optics, and
Newton's to mechanics are still regarded as models of the axiomatic method generally. The others, lacking in rigor by recent
standards, offer a challenge to the student interested in setting up
their contents more systematically.

Right at the start, however, Plato posed a problem whose two
main solutions guided and divided all subsequent scientists. This
is the problem of the nature and derivation of the axioms or basic
premises. Are they true descriptions of the world, or are they
merely hypotheses or calculating devices? If the axioms are true,
then the theorems are true. But what guarantees the truth of the
axioms? The axioms are not demonstrable in the way that what
follows from them is demonstrable. Is it therefore a case *first,* of
establishing the truth of the axioms in some other way, and then,
afterwards, of setting up the true system as economically as possible? Or is it a case of *freely inventing* those axioms that will most
economically account for the facts?

In presenting the two solutions Plato gave his version of the
nature of scientific method.[3] He enriched his account with a host

3. *Republic,* 509-11, 533-34.

of metaphors drawn from space-relations, travel, cave-dwelling, hand-grasping, music, and architecture. He conceived the whole structure of the correct solution as a stepped arch one side of which the scientist must mount before he descends the other side.

In the incorrect solution the technician—according to Plato he is not really a scientist—merely descends part of one side of the arch, leaving the whole structure hanging in the air. That is to say, he begins and completes a deductive argument without having first established the truth of his premises. This is the way of geometers who invent hypotheses about figures and angles and other such data, and with these proceed forthwith to set up a deductive system. "Having adopted their hypotheses, they decline to give any account of them, either to themselves or to others, on the assumption that they are self-evident. Then, starting from these assumptions, they descend deductively by a series of consistent steps, until they arrive at all the conclusions they set out to investigate." This procedure is certainly useful in technical matters: it works. Because it works in all phases of earth-measuring this procedure is perfectly accommodated to the purpose of the geometers. But their purpose is wrong. "Geometers constantly talk as if their object were *to do,* whereas the true purpose of the whole subject is *to know.*" If this is so, then the procedure fails. How can we know that the conclusions are true unless we know that the premises are true? They are not known to be true because "the geometers leave the hypotheses they use unexamined." In which case, "if their premises are things they do not really know, and their conclusions and the intermediate steps are deduced from things they do not really know, their reasoning may be self-consistent, but how can it produce science [*episteme*]?"

The correct solution avoids this mistake. This is because the scientist starts from the bottom of the other side of the arch. In the upward journey he "treats his assumptions, not as first principles, but as *hypotheses* in the literal sense—things 'laid down' like a flight of steps up which he may mount all the way to something that is not hypothetical [in the metaphorical sense]." Not until this is done can the scientist proceed to set up or, rather, lay down, his system. Not until he has grasped certain truths (for Plato, it should be noted, there is one that is ultimate) can he use them as first principles to start his demonstration. The scientist "turns back," and through a series of consistent steps, "descends at last to a conclusion" by deduction. Not until all this is done, do the special skills of arithmetic, geometry, astronomy, harmonics, etc., become sciences. They take their proper places within the structure of the arch. Their "first principles," so-called, are used as steps to knowledge of the things at the top of the arch, the real *archai* or beginnings of demonstration. Thus is achieved a *hierarchy,* in at least two senses, of the sciences, for their "first principles," previously used as steps in the ascent, may become theorems in the descent. What once were hypothetical, now, "when connected with a first principle, become intelligible."

The upward journey has two main stages. The first or preliminary stage consists in engaging in all the special skills and in becoming clever at them. Here Plato mixed in another metaphor. Engaging in these skills is only a "prelude to the main melody" called the dialectic. Invented by Socrates, it amounts to the proposal of hypotheses which are then tested, "done away with," and "converted" into truths by induction. Corresponding to this last step in the inductive procedure is the *mental state* present when we "grasp" truth. It is called intuition. Among the hypotheses proposed for testing are the "first principles" of the special skills previously used without question. It is the dialectic, therefore, that enables us to convert the practical arts or skills into sciences. Accordingly, "dialectic will stand as the coping-stone of the whole structure."

Reducing Plato's account to terms of analysis and synthesis: In the incorrect solution the synthesis proceeds without previous analysis. This entails deduction without demonstration. In the correct solution, analysis must precede synthesis, that is, there should be no synthesis without previous analysis; and in Plato's own doctrine, all the stress is upon analysis. This enables two sorts of things to become intelligible: the first principles themselves, and the theorems deductively derived from them.

The methods of subsequent scientists are illuminated by treating them as variants upon Plato's original. From his point of view their users make two classes: those who really are scientists, and those who are technicians. Before treating Descartes and Newton, it will be helpful to consider briefly two earlier followers of Plato, namely, Aristotle and Euclid, the one a "scientist," the other a "technician."

Aristotle was a "scientist" because he adopted Plato's distinction between analysis and synthesis. But he used a different illustration—that of a race-course: "There is a difference between arguments *from* and those *to* the first principles. For Plato, too, was right in raising this question and asking, as he used to do, 'Are we on the way from or to the first principles?' There is a difference as there is in a race-course between the course from the judges to the turning point and the way back."[4] Moreover, he laid all the stress upon analysis—the arguments *to* the first principles—and he accepted the doctrine that there should be no synthesis without previous analysis. That is, he held that the premises of science must be principles: they must be known to be true before demonstration. This is true even of hypotheses, for it was left to a later age to make hypotheses hypothetical. He followed his teacher, though with added detail, in his account of the derivation of premises by the method of induction—now known as scientific discovery—and he too isolated that mental state corresponding to the last stage of the inductive procedure by which we apprehend truth, intuition. Nevertheless, he clarified Plato's distinction between knowledge of a principle and knowledge of what is "connected with a principle." Only the latter properly belongs to science. That is to say, all science is synthesis or demonstration. It is knowledge not of the premises but of what comes after the premises. But it rests upon the previously acquired "more accurate" unscientific knowledge just mentioned. All this was more tersely put by Aristotle himself in four assertions:[5] "We come to grasp the first principles only through [the *method* of] induction"; "it is [the *mental state* of] intuition that apprehends the first principles"; "Scientific knowing and intuition are always true"; and "All scientific knowledge is discursive."

Euclid, on the other hand, was a "technician" because his *Elements* satisfy the proscribed doctrine of synthesis without previous analysis. His common notions and postulates are all hypotheses

4. *Nicomachean Ethics*, 1095a.
5. *Posterior Analytics*, 81b, 100b.

because he "declined to give any account of them either to himself or to others on the ground that they were self-evident" and proceeded forthwith to set up his deductive system, thus leaving the structure of his argument hanging in the air. But these remarks fit Euclid's works as they now stand. Perhaps he left another book, now lost, in which he showed how he had derived his postulates from higher principles like those attempts made later by Ptolemy, Proclus, Nasiraddin at-Tusi, and Legendre to prove Euclid's own fifth postulate. Or, more likely, he left a book, also lost, written in the same analytical style as Newton's *Opticks,* in which he showed how he had derived his postulates by induction. Or, most likely, he thought that all his premises were true for good reasons but, preferring to write only in the synthetical style, never showed how he had derived them. Considerations like these prompted a later geometer to conclude:

> It was this synthesis alone that the ancient geometers used in their writings, not because they were wholly ignorant of the analytic method but, in my opinion, because they set so high a value on it that they wished to keep it to themselves as an important secret.[6]

The geometer was Descartes. The "important secret" he tried to reveal.

3. Descartes

Both Descartes and Newton were Platonists in their adoption of the doctrine that analysis precedes synthesis. Descartes wrote:

> It is certain that, in order to discover truth, we should always begin with particular notions in order to reach general notions afterwards, though reciprocally, after having discovered the general notions, we may deduce from them others which are particular.[7]

6. *Second Replies.*
7. Letter to Clerselier.

Newton was just as explicit:

As in mathematics, so in natural philosophy, the investigation of difficult things by the method of analysis ought ever to precede the method of composition.[8]

In the matter of the ingredients of the two procedures, however, they differed from Plato and between themselves. But even here there is enough resemblance to enable us to see the two later accounts as different interpretations of the one original.

Descartes was primarily a physicist. Although his *Discourse on Method, Geometry,* and *Meditations on First Philosophy* have had tremendous influence in their respective fields, his main interest lay neither in method, nor in abstract geometry, nor in metaphysics but in the application of these subjects to physics. He "designed to devote all his life" to the discovery of "a practical philosophy" which, replacing the speculative philosophy of the Schools, would make ourselves masters and owners of nature.[9] Accordingly, the pattern of his life was made up of the early discovery of the method, a general science, restricted to no special subject, and called *"Mathesis Universalis"*; then its application to geometry, optics, and astronomy; and finally the discovery of the foundations of his physics. Throughout all this, however, he was doing nothing but geometry. His method, devised to solve "problems about order and measurement," was geometrical. The special subjects to which he applied it were geometrical. And when he gave up abstract geometry he said: "I am doing this purposely in order to have more time to study another kind of geometry," for "my physics are nothing else but geometry." [10]

If metaphysics was secondary, what was its specific role in his entire scheme? It was to provide the foundations of physics. Descartes' quest was for certainty. He could get certainty in physics only by getting true premises. And these could be true only if they were grounded in metaphysics. The search for the foundations ended in 1641, only nine years before his death.

8. *Opticks* (1704), Query 31.
9. *Discourse,* Part Six.
10. Letter to Mersenne (1638).

Having finished his *Meditations,* he privately admitted: "I shall tell you between ourselves that these six meditations contain the entire foundations of my physics."[11] Why were these so relevant? The search had been long. Evidently his first venture into physics, *The World,* begun in 1629 and never published, had been composed without satisfactory foundations. As late as 1638 he had said that it is impossible to provide a demonstration of matters that depend on physics "without having proved the principles of physics previously by metaphysics," and he implied that neither he nor anyone before him had done so.[12] In 1640, however, the "break-through" was imminent, for he said, "having reduced physics to the laws of mathematics, the demonstration is now possible."[13] But still this was not enough because the premises might not be true. Their referents, material objects, might not exist. "I already know," he wrote at the start of the sixth meditation, "the *possibility* of their existence in so far as they are the subject-matter of pure mathematics since I clearly and distinctly perceive them." But for the same reason that an atheist can infer he is awake but can never be certain,[14] so he cannot be certain of the existence of material objects and thus of the premises of physics. This certainty must come from God the non-deceiver, and His certainty must come from metaphysical proof. This the *Meditations* provided. Accordingly, the things clearly and distinctly perceived exist. Therefore the objects of solid geometry exist, that is things extended in length, breadth, and depth. And these are nothing but the referents of the premises of physics. Thus Descartes reached the end of his search, the beginning of demonstration. This was the ultimate justification of his method.

In one passage in Part Six of his *Discourse* Descartes revealed the working sequence of his method. This I divide into two parts:

> My general order of procedure has been this. First I have tried to discover in general the principles or first causes of all that exists or could exist in the world. To this end I con-

11. Letter to Mersenne (1641).
12. Letter to Mersenne (1638).
13. Letter to Mersenne (1640).
14. *Third Replies.* The objector was Hobbes.

sider only God, who created them, and I derive them merely from certain root-truths that occur naturally to our minds.

Then I consider the first and most ordinary effects deducible from these causes, and it seems to me that in this way I discovered the heavens, the stars, an earth, and even on the earth, water, air, fire, the minerals, and some other such things. . . . And then I tried to descend to more special cases. But in view of their wide variety I thought it impossible to distinguish those actually found on earth from those that could be found there. . . . It thus appeared impossible to proceed further deductively, and if we were to understand and make use of these things, we should have to discover causes by their effects, and make use of many experiments. . . . My greatest difficulty usually is to find out which is the true explanation, and to do this I know no other way than to seek several experiments such that their outcomes would be different according to the choice of hypothesis.

This is Descartes' version of the two Greek procedures of analysis and synthesis illustrated by Plato as the ascent and the descent of a stepped arch, and even similarly illustrated by Descartes as "ascending in steps" and "descending," respectively. Now the first strange fact about his account of method is his complete neglect in all the rules specified in the *Regulae* and the *Discourse* of the method of synthesis (characterized in the second part of the passage). What explains it? His interest—like Plato's and unlike Euclid's—lay less in system building than in scientific discovery. And this was because, while the former was so well known and easy and ought to be transferred to rhetoric, the latter was unknown and difficult, Euclid and others having "grudged the secret to posterity." The one "does not teach the fashion in which the matter in question was discovered"; the other does.[15]

Yet the details of his analysis are much the same as Plato's. In the quoted passage (first part) he said that from certain "root-truths" we derive "the principles." But this describes only the second of the two movements, both of which he gave in Rule V of

15. *Regulae,* IV and X; *Second Replies.*

the *Regulae:* "We reduce complex and obscure propositions step by step to simpler ones, *and then,* by retracing our steps, try to rise by intuition of all the simplest ones to knowledge of all the rest"; and which he repeated in the four specified rules of the *Discourse:* We "divide into parts," accepting as true only what is "clearly and distinctly" presented to the mind, *and then* "ascend little by little, in steps, as it were, to the knowledge of the most complex." Thus within the analysis itself there are two movements: the "descent," involving the famous method of doubt which, after a series of deductive steps, culminates in intuitions of certainties, such as our own existence and the nature of a triangle; and the "ascent" from these, by a series of intuitions, to such truths as the laws of motion which may become principles. The two dominant "intellectual activities" here, both of which originate from "the light of reason," are intuition or indubitable conceiving, and deduction, which is the same thing applied to more than one step of an argument. The whole analysis then, begins with "complex and obscure" propositions and ends with only "complex" ones, the obscurity having been dispelled. This is much the same as Plato's dialectic which begins with hypotheses and ends with their conversion into known truths by intuition, their hypothetical content having been eliminated. But by trying to transfer the certainty of geometrical demonstration to the procedure of scientific discovery, that is, the certainty of synthesis to analysis, Descartes thought he had found the secret of the ancient geometers.

His account of the synthesis, however, contains a startling disclosure. Having laid down his rules of method, Descartes proceeded to deviate from them. We expect a demonstration of the truth *after* the principles have already been found, such as had been prescribed by Plato. But we get a different way of accounting for the facts: "the way of hypothesis" proscribed by Plato. Unable to proceed further deductively, he resorted to the invention of and choice among different hypotheses, the choice being determined by crucial experiments. Thus he gave up the certainty of the *a priori* method in favor of the conjectural *a posteriori.* This meant that in his practice the synthesis preceded analysis, for it preceded that form of it known as inductive testing. The second half of the quoted passage describes this deviation. His actual work in physics and optics reflects it.

In the third part of the *Principles* he said: "I will put forward everything that I am going to write just as a hypothesis. Even if this be thought to be false, I shall think my achievement worth while if all inferences from it agree with experience"; and "I shall also here assume some propositions which are agreed to be false"; to which he added that "the falsity of these propositions does not prevent what may be deduced from them from being true." The hypothesis in question, contrary to the Church doctrine of full creation, was that the matter of the universe originally consisted of small particles of the same shape each moving round its own center; or, in other words, as he said in the fourth part: "I have described the Earth and the whole visible universe *as if it were* a machine, having regard only to the shape and movement of its parts."

Nevertheless, it seems that all Descartes' make-believe was only make-believe. Although his big hypothesis may have been freely invented, there is little doubt that he thought it true. For he preceded his introduction of hypotheses in the *Principles* with the remark: "It can hardly be otherwise than that the principles from which all phenomena are clearly deduced are true." In the sixth part of the *Discourse,* with regard to the hypotheses he had used in optics and astronomy, he said that "the truth of the hypotheses is proved by the actuality of the effects," and "I think I can deduce them from the first truths." Moreover, it should be noted that the content of his big hypothesis is the physical world shorn merely of the secondary qualities of color, smell, taste, and so on. It is extended matter, the defining feature of the physical world, the long-sought after and scrupulously argued for conclusion of all the *Meditations.* And this is nothing but the object of solid geometry which, with the addition of another primitive, motion, makes it the object of physics. It would be queer to suppose that Descartes thought his hypothesis false. Finally it should be noted that this "hypothesis" of extension plus motion was used in conjunction with some things not hypothetical, namely, the three "laws of nature" previously deduced. Given extension and motion these laws of motion enabled Descartes to construct the world.

Why then did Descartes use the word "hypothesis"? Its use, fairly obviously, was dictated by prudence. Now in 1633 he had suppressed his *The World* after hearing of Galileo's censure by the church. This book showed how a world would inevitably form from the laws of motion operating upon an original chaos. But it must not be thought that he suppressed it because the chaos was not expressly hypothetical. It was. The "hypothetical" chaos fully satisfied the injunction of Cardinal Bellarmine, the inquisitor of Bruno: "Galileo will act prudently if he will speak hypothetically [*ex suppositione*]." He suppressed it partly because in it he had maintained the motion round the sun not only of the "hypothetical" earth but, like Galileo, of the real earth; and partly because "there might be," he said, "other opinions of mine in which I was misled." It seems likely that, long before all this, he had decided to adopt "the way of hypothesis" in expounding his physics. He had decided to "*speak* hypothetically" though not to think in that way.

4. Newton

We are greatly helped in our understanding of Newton's method by seeing it not only as a variation within the classical tradition but as a reaction against the method of Descartes. In the preface of the *Principia* Newton showed the direction he was taking: "The whole burden of philosophy seems to consist in this —from the phenomena of motions to investigate the forces of nature, and then from these forces to demonstrate the other phenomena." Thus, like Descartes, he accepted the Platonic distinction between analysis and synthesis as well as the Platonic sequence in which the analysis comes first. Nevertheless, he corrected Descartes in two main ways. First, like Plato, he rejected "the way of hypothesis." Secondly, although, like Plato and Descartes, he stressed the way of analysis, he restored one of its defining features: the conclusions of analysis are discovered not by intuition independently of experience but by experiment and observation. He thus rejected Descartes' "The Light of Reason" in favor of his own "The Light of Nature."

But Newton's extraordinary rejection of hypotheses from physics and his denial that he used them create an initial problem in understanding him. In the second edition (1713) of the *Principia* he claimed: *"Hypotheses non fingo"*—literally, "I do not invent (or frame or feign) hypotheses"—and in the *Opticks* he wrote: "Hypotheses are not to be regarded in experimental philosophy." These are strange claims, for Newton did frame hypotheses time out of mind. The *Opticks* indeed is riddled with hypotheses about the nature of light, the existence of aether, the nature of the first cause, and so on. From this apparent discrepancy it would be easy to conclude that Newton said one thing and did another. But it is unlikely that this is the correct answer. How are we to reconcile what he said he did with what he did? We are hindered rather than helped by his careless negative definition of "hypothesis": "Whatever is not deducible from phenomena," which includes almost everything. We need, then, to turn to the context in which he rejected hypotheses, specifically to the rules of method that he set down at the end of his *Opticks* long after he had applied them.

The method of analysis ought ever to precede the method of composition. This analysis consists in making experiments and observations, and in drawing general conclusions from them by induction, and of admitting no objections against the conclusions but such as are taken from experiments or other certain truths. For hypotheses are not to be regarded in experimental philosophy. . . . By this way of analysis we may proceed . . . from effects to their causes. . . . And the synthesis consists in assuming the causes discovered and established as principles, and by them explaining the phenomena proceeding from them and proving the explanations.

It is clear from this remarkable passage, essential to the understanding of Newton, what he was advocating. He was advocating no synthesis without previous analysis from experience, that is, no use of the axiomatic method without having first derived the axioms from experience by induction. It is equally clear what he was rejecting. He was rejecting "the way of hypothesis," which is nothing but the method of Descartes: synthesis without previous analysis was what Descartes *said* he practised, while analysis without experience was what he preached. The former is actually "the way of hypothesis" or the way of *inventing* the premises to start a demonstration, but the latter also involves the same way, at least from Newton's point of view, because the premises are not derived from experience.

Accordingly, when Newton said, "Hypotheses are not to be regarded," he meant simply that induction from experience ought to precede deduction. When he uttered his *"Hypotheses non fingo"* he was saying in a very abbreviated, and hence cryptic, way: In induction, I do not invent hypotheses, and in deduction I do not demonstrate from them. More fully, he meant that the inductive side of scientific method has a beginning, a middle, and an end, and all must be complete before any deductive system is set up. The beginning consists in "hinting several things" or making "conjectures" about the causes of phenomena. Such hinting is consistent with "I do not invent hypotheses" because they are "plausible consequences" drawn from the facts. That is, they are not derived, like Descartes' conclusions, merely by the Light of Reason or intuition. Although hypothetical in character, Newton did not call them "hypotheses." The middle consists of examining these "hints" and improving them by observations and the tests of experiment. The end is defined by his remark: "And if no exception occur from phenomena, the conclusion may be pronounced generally" and considered "proved" as a "general law of nature." *"Afterwards,"* the deduction proceeds by assuming the conclusions established as principles, and from them demonstrating the phenomena. This procedure is consistent with "I do not demonstrate from hypotheses" because the first principles, although they are *assumed* true, are grounded in experience. The peculiar character of this method, the stress upon experience and the rejection of hypotheses of the Cartesian kind, may be briefly described as follows in Berkeley's words: "It is one thing to arrive at general laws of nature from a contemplation of the phenomena, and another to frame an hypothesis, and from thence deduce the phenomena *(S, 229)*." Had Newton said this, his code words *"Hypotheses non fingo"* would need little deciphering. The characteristic example of the sort of hypothesis proscribed in this view is the epicycle.

Let us see whether the explanation of Newton's meaning just given saves the appearances, where the appearances are Newton's actual applications of method. The *Opticks* is the best illustration of Newton's method, for it not only contains every feature but the stress is where he wanted it, on induction. Structurally this is a very strange work. It begins with "definitions" and "axioms." The axioms, Newton said, "I content myself to assume under the notion of principles in order to what I have farther to write." Then follow "propositions," "theorems," and their "proofs." All this suggests that the *Opticks* is a work of synthesis, a deductive system after the manner of Euclid's *Optics*. But it is not. None of the theorems follows from the axioms. Each proof, except for three at the end of Book I, is a "proof by experiments" although the axioms are assumed in these proofs. Accordingly, the *Opticks* is almost entirely an illustration of the inductive procedure. Books I and II show how Newton derived laws of nature by experiment and observation. Here the book is like a cook book written by an expert chef long after all the recipes have been tried. In a typical "proof" there is an autobiographical account of experiments actually performed, thus giving a recipe that anyone can use if he has the ingredients. From the experiments, the laws of nature are "manifest." In many cases Newton contented himself merely with recording "observations." In Book III, on the other hand, Newton had "only begun the analysis." It contains what we should call "hypotheses" framed as twenty-nine "queries" and two "questions." While, in the first two books, certain "manifest qualities" of light, such as that white light is a compound of all the primary colors had been "proved," here queries are proposed for further research by others.

The queries are a revelation of an extraordinary mind, disciplined yet free, marvellously keen on detail, yet bold in imagination. These negative queries are really positive conclusions of arguments that are all empirical, all going from effect to cause, and all inductive. The conclusions, however, are nothing but hypotheses which, more optical and less speculative in the first edition, become increasingly less optical and more hypothetical in later editions. This third book of the *Opticks* is not only a commentary on optics; it is a commentary on physics, a work in empirical metaphysics, and a work in empirical theology—all of which, Newton held, are legitimately included in physics. Its premises, mainly implicit in this part, are drawn from optical and other physical phenomena, and the ultimate conclusion is God. In 1704, Newton argued for such things as the mutual interaction of bodies and light, and for the existence of vibrations in the retina, the effect of light and the cause of sight. In 1706, he argued for the existence of aether, a new medium "swifter than light," and "exceedingly more rare and subtile than the air," and once again for the existence of vibrations in the retina, this time chiefly the effect of aether. In 1717, he speculated that light is made of small unbreakable bodies, and he made conjectures about the existence and nature of God: "Was the eye contrived without skill in optics . . . ," and "Does it not appear from phenomena that there is a being incorporeal, living, intelligent, omnipresent . . ." who alone is directly acquainted with physical objects, while mortals perceive only their images? All these conjectures were "plausible consequences" of the empirical procedure, but since they were not yet "proved," they were not yet ready for use as premises in demonstration.

Because induction must precede deduction, the *Opticks* was a preparation for a demonstration that Newton never offered. That he intended to set up a deductive system of optics is suggested by the fact that the whole of the *Opticks* is styled "Part I." From this account we see that the *Opticks* is perfectly consistent with *"Hypotheses non fingo"*: all the "hypotheses," Newton would have argued, are conclusions from induction, and none are yet to be used as premises in a deduction.

In the *Principia,* however, Newton's good resolutions broke down, for there is one disconfirming instance of the explanation I have offered. In the first edition at the beginning of Book III Newton had presented nine "hypotheses." In the second edition (1713), in line with his rejection of hypotheses, he transformed all of them into "rules," not used in demonstration, with one exception which he retained as "Hypothesis I: That the centre of the system of the world is immovable." From this he proceeded to demonstrate. This, surely, is a case of *hypotheses fingere.* But, after stating it, Newton said immediately: "This is acknowledged by all." This remark echoes what he had said about the axioms of his *Opticks* in 1704: "What *has been generally agreed on* I content myself to assume under the notion of principles." He was doing the same thing here. But whereas the optical axioms might be shown to appear from phenomena, it would be difficult to do the same with this hypothesis. Nevertheless, Newton evidently had to use it, and, accordingly, broke his resolution *"Hypotheses non fingo."*

The rest of the *Principia* does jibe with what he said about method. Force, gravity, and attraction are used here as principles in demonstration. It is clear from question 31 of the *Opticks* that Newton thought that none of them are "occult qualities" or hypotheses. They are all consequences from analysis, "active principles," "manifest qualities," "general laws of nature," or "causes" of motion. While the causes of motion had been discovered, *their* cause had not. Nevertheless, consistent with his rejection of hypotheses, Newton was able to speculate about their cause, that is, to make the hypothesis in the General Scholium to Book III of the *Principia* that it was "a certain most subtle spirit," namely, "aether." In the same place he also speculated about the nature of God, "to discourse of whom" is a proper part of physics. Again it should be noted, however, that he used neither aether nor God as a principle of demonstration.

The main features of Newton's method, it seems, are: the rejection of hypotheses, the stress upon induction, the working sequence (induction precedes deduction), and the inclusion of metaphysical arguments in phyics. For final confirmation that Newton so regarded his own method I refer to query 28 of the *Opticks* where he appealed to the authority of the best tradition of the Greeks against the fashion of "late philosophers" who, "feigning hypotheses for explaining all things mechanically and referring other causes to metaphysics," banish non-mechanical causes from physics. Then he summed up his own view: "The main business of natural philosophy is to argue from phenomena without feigning hypotheses, and to deduce causes from effects till we come to the very first cause which certainly is not mechanical."

5. "The Sciences now have Masks on them"

Having extracted the features of their methods that I need, I now begin to show how Descartes and Newton were victimized by their metaphors, victimized because they presented the facts of one sort as if they belonged to another, but without awareness. They were engaged in sort-crossing. But because they did not know that they were, they confused their own peculiar sorting of the facts with the facts. It was as though, having found that wolf-properties were eminently suitable for illustrating man, they came to believe that he was indeed a wolf. Of their many sort-crossings I shall isolate three.

The first is that of the deductive relation with the relation between events. The former relation belongs to procedure. It is, therefore, the sort of thing that is invented; it was, in fact, invented by the Greeks as the best way of teaching. The latter relation belongs to the process going on in nature. It is, therefore, the sort of thing that is discovered. Now Descartes and Newton adopted the deductive procedure as a most powerful instrument. It was a defining feature of their *"more geometrico"* and "mathematical way," respectively. Descartes' "long chains of reasoning" were deductively linked. Newton's demonstrations were reduced to "the form of propositions in the mathematical way." These ways involved, as we have seen, the deduction of conclusions (theorems) from premises (principles). According to Descartes, "we deduce an account of the effects from the causes." According to Newton "the synthesis consists in assuming the causes discovered and established as principles, and by them explaining the phenomena." It is clear that both men thought that principle and theorem were necessarily connected—necessarily, because this had been a matter of decision by their teachers, the Greeks. All this, it was decided, occurs in the procedure.

What did they think they found, however, going on in the process? The answer is astonishing. The physicist's procedure is duplicated in the physical process. The principle of procedure that starts a demonstration is repeated in the "active principle" that starts a causal process. Moreover, the relation between a principle of procedure and its deduced consequences is exactly the same as that between an "active principle," such as gravity, and its effects.

This relation is that of necessary connection. Both men thought that physical causes produce the existence of their effects, and that the effects necessarily follow from the causes, for Newton described the effects as "proceeding from them," and Descartes supposed that all the phenomena now in the world would be produced by necessary consequence from the laws of motion acting upon either the chaos of the poets or the originally ordered extended particles.

Just to set down this still familiar view is enough to show its strangeness and enough to suggest the aetiology of the confusion involved. The law of nature that explains it is the Principle of Association according to which things noticed constantly to go together first suggest each other, then get the same names, and finally come to be thought the same thing or of the same sort. Since the time of Euclid, physical process and the deductive procedure used to explain it have been constantly associated in the minds of scientists. Thus a procedure-process shift has occurred, at first one suggesting the other, then both sharing the same names: "principle," "necessary connection," "necessary consequence," "proceeding from," and "system"; but then, inexorably, becoming things of the same sort. In this fashion a defining feature of deductive argument was exported to the external world—a prominent page of the recipe was mixed in with the stew. Nature, it was concluded, obeys the logic of the deductive method.

The second sort-crossing I exhibit is the inadvertent identification of explanation with physical explanation and this with causal explanation, that is, the reduction of one to the other. The main elements of this confusion appear from the account just given. Here are additional details. According to Newton the main task of physics was to find the forces of nature, "and then from these forces to demonstrate the other phenomena," the other phenomena being bodily motions. Descartes' view, although earlier in time, was an advance on this, for he saw the task as that of finding the laws of these motions and using the laws to demonstrate the motions. But both men thought the explanation had to be causal, the one holding that physical forces, the other that physical laws, cause events. Moreover, Newton's three "laws of motion" and Descartes' three "natural laws" reveal that the ultimate constituents of their world were of two sorts: the effects, such as "bodies moving" and "bodies at rest"; and the causes, such as "power of going," "external causes," and "resistance," in the case of Descartes, and "impressed forces," in the case of Newton, specifically described elsewhere as "gravity" and "attraction." These entities were described either as "clear and distinct ideas" or as "manifest qualities" to distinguish them from "obscure notions" or "occult qualities." It should be noted once more that the word "principle" was used to refer ambiguously: to the premise or statement of the law in the procedure, and, under the name "active principle," to the supposed cause in the process.

Now although no one since the time of Descartes and Newton has been able to find these forces or active principles in the physical world, many have thought they were investigating them. What is the aetiology of this prejudice? It is easy to accept that corollary of Newton's first law of motion, itself a corollary of Descartes' first law, that every body alters its state only if compelled by impressed forces. But this law has its psychological foundation in another very simple law known by us much more intimately: We are aware that we can make and do things, push and pull them, act with and react to them; we are aware of ourselves as agents, causes, forces, or minds; and this awareness comes to us in infancy and remains as secure as ever. Here, it seems, lies the origin of all our notions of cause. It is often associated with movement, the movement of our limbs and muscles, though it need not be, for we can decide to think of giraffes or kangaroos without the flicker of an eyelid. At an early age we make three inferences: (1) there are other agents or forces beside ourselves; (2) every body alters its state only if forced; (3) bodies are or contain forces. Many of us retain such beliefs throughout our lives. We ought, however, to be suspicious of, at least, the third, because it seems impossible to find the force or power referred to. The only actual experience I have of power, force, or activity is in myself. Outside myself I notice events following each other in time, and I find it very hard to resist the tendency to ascribe power or activity to many of the earlier events. When primitive societies ascribe power to clouds, mountains, and rocks, we are amused. When Descartes and Newton and their followers ascribe force to bodies, it seems to make sense. But surely both are cases of what might be called hylopsychism, because in both something that we find only in ourselves, something that belongs to persons or living things, is ascribed to matter.

It would be a mistake, however, to conclude from this that efficient causes are to be excluded from science. For this would be to equate science with physical science, which only someone steeped in the parochialism of a particular age would want to do. From the analysis of what it is to be a cause, I conclude that there are efficient causes. This allows us to have a science of persons, for we have knowledge of them, and, if so, we can systematize it. Such efficient causes, then, are to be admitted as objects into the principles of science but are to be excluded from the principles of physics.

The third case of sort-crossing I want to exhibit is the unwarranted identification of deduction with computation or calculation or any other form of metrical reckoning or counting. This confusion accompanies a narrow view of the nature of science and scientific demonstration. For some have so narrowly conceived the Cartesian "geometrical method" or the Newtonian "mathematical way," that they (including the authors) have not realized that this method or way—paradoxical as it may seem—need be neither geometrical nor mathematical. Its defining property is demonstration, not the nature of the terms used in it. Whether these terms are used to denote measurable qualities, lines, or angles, or the god's portion of the anatomy of wallaroos is accidental to demonstration itself, and thus accidental to science. What is the aetiology of this prejudice? Measurable qualities, treated mathematically, are used with great advantage in the sciences, but, once more, we must beware of the wide subliminal applications of that law of nature by which frequent associations —especially successful ones—are confused with identity. Because mathematical computation is constantly used in science, we must not regard it as a defining property. Because lines and angles are used to enormous advantage in optical demonstration, and have been so used constantly since Euclid's time, we must not therefore succumb to the tendency to think that explanation by means of lines and angles exhausts optical explanation. We might just as well say that mechanical explanation exhausts science or that we cannot set up a deductive system without using differential equations.

6. "If the Masks were taken off . . ."

These confusions, because they involve sort-crossing without awareness, seem to be nothing but disguises or masks placed by men on the faces of nature and procedure. It seems, then, that all we have to do is to lift them or take them off in order to see what really occurs and what the scientist actually does. Accordingly, necessary connection illegitimately held to obtain between the events occurring in the process of nature, is restored to its proper place as an ingredient only of procedure. Moreover, when it is said that motion and change in the world are caused by attraction or other active principles, it seems likely that all that is meant is that bodies move in a certain order. Accordingly, the "active principles," star performers in process, are restored to their proper role as principles only in procedure. The principles of physics or any science are premises from which other statements are deduced that describe the way events unfold. The phrase "deduce effects from causes" is simply translated in terms of premises and conclusions, and causal explanations are replaced by explanations in terms of laws or rules; rules because, having lifted the disguise from the face of nature, all we find are bodies moving in a certain order, that is, regularly, that is, according to rule. Thus the causal statement "x attracts y" is replaced by "x and y move according to rule"; and similarly for a thousand similar statements. Finally, the deductive procedure is defined as a special relation between symbols whose interpretation, either numerical or not, is merely accidental.

What I have been doing resembles what Socrates did in the *Phaedo* in his rejoinder to the argument of Cebes. When young he was enthralled by that wisdom called Physical Science, and was "completely blinded" by these studies. He fell for the "mechanism" of the times and identified explanation with mechanical explanation, cause with physical cause. With great difficulty he extricated himself from the use of this appealing procedure and invented a new one. He replaced explanation in terms of causes and effects by explanation in terms of reasons and their deduced consequences. For example, he replaced the mechanical explanation of his "sitting here now," in which nothing was considered but the parts of his body and their motions, by explanation in terms of reasons. He rejected physical causes as causes at all. "To call these things causes is too absurd." His own advance, however, he spoiled by assigning causal power to his reasons.

7. Replacing the Masks

But there was something inadequate and naïve about the attempt made in the last section. It was overambitious. Who am I to say what is the correct sorting as though the sorts, having been made in Heaven, were then laid out for us to observe? Moreover, the methods I have been describing worked in spite of their authors' beliefs about them. It would be more sophisticated, therefore, to leave the sort-crossing as it was, except for this important difference: We have now become aware of it. It is, then, as though we now agree with Descartes and Newton that the wolf-properties are peculiarly suited to illustrate man. But, unlike them, we merely refrain from being taken in by this device.

This brings me back to the subject with which I began this chapter, that of hypotheses. Plato, we saw, extended the metaphor in the word "hypothesis" in order to assign to hypotheses their proper role. They were to be treated not as *archai* to start the demonstration (things at the top of the arch) but as *hypo-theses* to start the analysis (steps at the bottom of the arch) from which, after their hypothetical content had been destroyed, the *archai* might be derived. I concluded that Descartes and Newton adopted much of this view, for they too held that we ought to start the analysis with hypotheses, the "obscure propositions" of Descartes and the "hints" or "conjectures" of Newton, and finish it with those established truths called laws of nature, which can function as first principles to start the demonstration. Both were really advocates of the Platonic view that hypotheses were not to be used in demonstration. Newton asserted *"Hypotheses non fingo"* and meant it. Descartes asserted the equivalent of *"Hypotheses fingo"* but did not mean it. He acted prudently by deciding to *"speak hypothetically"* while continuing, in line with his formal rules of method, to *believe* otherwise. This might well have been the "mask" he was referring to when he wrote the first of his *Private Thoughts:* "Now that I am to mount the stage of the world, where I have so far been a spectator, I come forward in a mask."

Nevertheless, in all their explanations of nature, both men were constantly making use of hypotheses. The "manifest qualities" that Newton found in bodies and that Descartes "clearly and distinctly" perceived in them were not discoveries of fact but were, from one point of view, occult qualities, and from another, free inventions of their highly imaginative minds. While they believed that they were giving true descriptions of the process of nature they were actually projecting the devices of one into the facts of the other.

Thus from one point of view (that of section 5), they were metaphysicians *malgré eux,* and their metaphysics was mechanism. But from another point of view (that of this section) they were inventing hypotheses, and their big hypothesis was that the whole physical world is one giant machine. Although Descartes himself even spoke from these two points of view, saying in the *Discourse* that "the laws of mechanics are *identical* with those of nature," and in the *Principles* that he had described the whole world *"as if* it were a machine," there is little doubt about which utterance he believed. The one implies belief, the other only make-believe. From the point of view attained in this section, both utterances reveal sort-crossing: the former, sort-trespassing; the latter, sort-crossing with awareness. From the same point of view, the former is a case of being victimized by a metaphor, while the latter is a case of using a metaphor. From the same point of view, the former characterizes the "reductionism" of physics to Euclidean geometry, while the latter characterizes either the extended application of geometry to the physical world or the physical interpretation of Euclid's geometry. Descartes and Newton gave slightly different meanings to "the geometrical way" and "the mathematical way," respectively. They invented slightly different hypotheses.

Moreover, within the classical tradition, both men thought that they derived their first principles by analysis, using the light of reason in one case and the light of nature in the other. Because they thought there was nothing hypothetical about them, they gave their results the dignified title of "laws of nature." Much of their written work they presented in the analytical style showing, they thought, the sequence of their investigation that ended in their discovery of these laws. They did not realize, it seems, that they were presenting, not their patterns of discovery, but detailed autobiographical accounts of their experiences undergone in testing their freely invented hypotheses. Einstein's discerning remark about Newton's *Opticks*, "It alone can afford us the enjoyment of a look at the personal activity of this unique man," applies just as appropriately to much of Descartes' written work. If this was so, if they first invented hypotheses and then subjected them to analysis in the form of inductive testing of their deduced consequences, then, while only one of them was following Cardinal Bellarmine's injunction to speak hypothetically, both were acting in that way. If this was so, then they were following the doctrine of the geometers, rejected by Plato, of setting up deductive systems based upon hypotheses or conjectures: They were "technicians." But if we consider Plato's masterpiece, the *Republic*, we must conclude that Plato himself, in actual practice, was following the same doctrine—the doctrine he rejected. For in it we see, not the derivation by analysis of first principles, but the arbitrary invention of one big hypothesis that is then used with extraordinary success to account for the facts about man.

Now all hypotheses, by definition, involve make-believe. Many of them, like those under consideration, involve sort-crossing, and are therefore metaphors. The conclusion of this section is the decision to try to adopt the actual technique of Plato and, in addition, to follow the advice of Cardinal Bellarmine. Then, whether we suppose that man is a state, or that the world is a machine, or that man is a wolf, the risk of confusing the facts of one sort with those of the other will be lessened.

CHAPTER III

New Metaphors for Old

1. Retrospect

THE CONCLUSIONS to be drawn from the preceding analysis will be more obvious if I reconstruct it at three levels. At the first level we notice that certain devices of procedure are confused with elements in the process. The process includes physical causes or agencies wearing a host of different guises such as gravity, attraction, repulsion, and force, that are necessarily connected with the effects they produce in nature. The procedure used to describe this includes taking these causes for principles, and from them deducing the effects. Since only measurable qualities are admitted into the principles, the deductive part of the procedure is identified with computation, which becomes a defining property of science.

But we notice, *first,* that these supposed agencies are probably only devices invented to describe and predict the events occurring. The corporeal forces or causes, not being observable, are occult. How did they get mixed up with the process? According to my analysis, force, a mere feeling, has been abstracted from persons and added to the physical world. Such hylopsychism is shared with primitive societies. It is nothing but a confusion of an extra- or meta-physical speculation, conjecture, or hypothesis with physics, and, therefore, a case of importing metaphysics into physics. *Secondly,* necessary connection, held to exist in the process, is again not observable. It is, therefore, occult. How did it get mixed up with the process? According to my analysis, necessary connection, traditionally the essential ingredient of the deductive procedure, has been abstracted from procedure and added to process. This categorial confusion is a most pedantic one. It is yet another case of importing an extra-physical quality into the physical world and, therefore, of importing metaphysics into physics. *Finally*, demonstration need not be mathematical. Not all sciences require differential equations. Symbols admitted into principles or premises need not be mathematical. The tendency to think that they must be is due to the fact that in geometry, the father of deductive science, and in optics and mechanics, its children, they always were. This is a case of parochialism in science and again of doing metaphysics without awareness, because it confers on the primary qualities, which are more readily treated quantitatively, a monopoly of existence.

At the second level of analysis I tried to show how the facts might be re-allocated to their proper places. Process and procedure are now sharply distinguished. Mechanical process in the physical world is reduced to bodies moving in a certain order, while procedure is reduced to premise and conclusion which, men have decided, are necessarily connected. But there are other things going on in the world beside bodily movements, such as tasting, loving, teaching, giving birth, praying, sacrificing, and waging war. These too can be systematically described by the same procedure. Accordingly, the interpretation given to the terms used in the procedure is treated as accidental to the procedure. If they can be given mathematical treatment, so much the better, but it is not a case of "joy and woe, being not yet measurable, there are no sciences of joy and woe."

At the third level of analysis apparently metaphysical features are promptly restored, but this time with awareness of their presence. The original methods have undergone metamorphosis but not a change of substance, for these methods worked in spite of their authors' beliefs about them. The authors were merely deluded in thinking they were doing one sort of thing when in fact they were doing another. For example, although I may believe that there are no such things as corporeal forces, I make believe that there are because such make-believe may be useful; that is, I feign hypotheses (*hypotheses fingo*).

2. Recognizing Hidden Metaphor

Let us get a different perspective on the preceding analysis. Consider it as the exposure of hidden metaphor. Corresponding to the three levels of analysis are three operations: first, the detection of the presence of the metaphor; second, the attempt to "undress" the metaphor by presenting the literal truth, "to behold the deformity of error we need only undress it"; and third, the restoration of the metaphor, only this time with awareness of its presence.

The successful performance of any one of these operations indicates that we have become aware of the presence of metaphor and, consequently, that we have avoided being victimized by it. If we are victimized, then we confuse devices of procedure with the actual process of nature, and thus unknowingly insinuate metaphysics. If we are not aware, then, like the duped citizens of the city-state of Oz who, owing to their green spectacles, thought everything was green, we add qualities that are the products of invention and decision, not of discovery. It is true that Descartes, when he invented his machine hypothesis, was the Wizard of Oz himself. But when Newton measured corporeal forces and found them to be "manifest qualities," he was just another duped citizen. Becoming aware of the presence of hidden metaphors in science, it dawns on us that there are other ways of viewing the world besides those that we inherit from the great sort-crossers of the past who, by their genius, hold us enthralled in just those attitudes that appealed to them.

For example, if the metaphor is one drawn from the physical to illustrate the mental, such as "the physical basis of mind," "the motion of the will," "mind is behavior," "the *id*," and so on, and we are captured by it, then we subscribe to a dogma that has no hope of salvation in it. Becoming aware of it, we start to expose psychohylism or behaviorism as one but not necessarily the only metaphysical theory of mind. If the metaphor used to illustrate nature is a mixed one drawn in part from something man has made like a clock or other machine, in part from something he did not make, like the feeling of force or power that we experience when we push or pull things, and we are enthralled by this dual picture, then we populate the world with a million ghosts in the giant clockwork of nature that lurk invisible, inaudible, and intangible behind its every movement. Becoming aware of it we start to expose mechanism as one but not necessarily the only metaphysics of nature.

The invention of a metaphor full of illustrative power is the achievement of genius. It is to create by saying "no" to the old associations, the things that have constantly gone together, the things already sorted, and "yes" to new associations by crossing old sorts to make new ones. But it is also an achievement to "undress" a hidden metaphor that has become part of the traditional way of allocating the facts, for this too involves breaking old associations. How is this to be done? How do we recognize a hidden metaphor? The question is much the same as the question: How do we avoid being victimized by metaphor so that instead of being used by it we use it? We can easily distinguish the mask from the face in many obvious examples of metaphor. We know at once that an airplane disaster may have been due to metal fatigue although metal did not grow weary; that famine, sword, and fire may crouch for employment without stooping; that sleep, which knits up the ravelled sleeve of care, achieves this without the help of knitting needles; that we can be tickled by the rub of love without giggling; and that death shall have no dominion without losing a throne. But how do we penetrate those disguises in which the make-up is hidden? The most common way enlarges upon what I have just been doing with obvious examples of metaphor. This I now illustrate and describe.

The burden of David Hume's refutation of the argument for the nature of God from the order or design found in the world amounts to the exposure of a metaphor, directly, by showing a weak analogy, and indirectly, by extending the metaphor. Consider the latter.[1] First, the great ship or the great house of the world has a "stupid mechanic" or builder who has "botched and bungled" many worlds before he struck out this one. Secondly, it has many builders who combined in its contrivance. Thirdly, the builders are male and female who propagate their species by generation. Finally, these anthropomorphic builders have eyes, a nose, mouth, ears, etc. By thus extending the metaphor, Hume

1. David Hume, *Dialogues concerning Natural Religion* (Edinburgh, 1779), Part V.

tries to reduce the argument to absurdity. The argument from design depends upon such metaphors as "The world is a ship or house," and "God is a builder." What Hume does is to focus attention upon the commonplaces associated with ships, houses, and builders, or, if these are to be considered as models, to extend the system of implications that has been adopted by such previous victims as St. Thomas and Sir Isaac Newton. That is, he adds more properties of the literal meanings of "ship," "house," and "human builder" to the world and God. This produces corollaries that he hopes are absurd.

But what are the mechanics of this absurdity? Precisely the same sort of thing is done when the metaphor involved in "Man is a wolf" is exposed. Let us suppose that St. Thomas and Sir Isaac Newton are again victims. We take "wolf" literally and transfer properties such as *four-legged* and *tailed* to man. We then ask the victims to test the wolf-hypothesis. Any man they meet is now a disconfirming instance, and it is hoped that they will reject the hypothesis. In the case of God who cannot be tested by observation to discover whether He is one or many, has eyes and nose, is married, and so on, what Hume intends is that we test it merely against our common notions of His attributes. The notion that He is a builder is acceptable, but that He is a woman is absurd.

The same operation is performed with ease and expedition when we make such conjunctions as "Men and timber-wolves are wolves," "The world and the *Queen Mary* had many builders," and "God and Frank Lloyd Wright are architects," because these are terse ways of extending the metaphor the whole way, or, rather, of treating two senses of a word, one of which is metaphorical, as one literal sense. By using such devices, which are nothing but abbreviated versions of Hume's, we expose the presence of metaphor and consequently avoid being victimized. It should be noted, however, that by extending the metaphor in part or completely to produce absurdity, we show only the presence of metaphor or sort-crossing. We do not show that something illegitimate, such as taking a metaphor literally or sort-trespassing is involved. It may have been the case that St. Thomas and Sir Isaac were fully aware that they were speaking in metaphor or were using models. In which case, like any user of models, they would have rejected any extension beyond the para-designer or the para-wolf hypothesis. Hume's method cuts no ice with those who are aware. It only reassures them.

This brings me to that peculiar case of metaphor known as myth. Some obscurity surrounds this concept. To start with, the expression "to explode a myth" is loose, because the metaphorical content, the double meaning, usually becomes apparent only to a later age. For those who make up the story or explanation or other device, which will later be called a myth, and for those who pass it on, there is usually no make-believe, only belief. The story is true, or the explanation gives the true ontology. There is only one meaning: the Theseus legend is no legend; the Minotaur is neither a fabulous monster nor a mythical one, he existed; the great white God depicted in the Australian corroboree actually came from the sea. To call a story or explanation a myth, however, is to accept it not as true but only as "a mendacious discourse figuring the truth."[2] It is to have recognized the story for what it is, a picture, not a true account. Trivial examples are old wives' tales. He who is the first to do this may make a new metaphor, for he may be the first to get the two meanings: the literal meaning of the legend, the story that is told, and the supposedly literal truth: for example, the nasty son of King Minos, the wooden bull, and the labyrinthine corridors in the Palace of Knossos; or, in the case of the corroboree, Captain Cook in his *Endeavour*. Thus unlike the fable, the parable, and the allegory, the story later to be called myth is usually seriously believed in by those who hear it. It is no metaphor. It is more than a likely story. It is more than the best explanation of the facts. It is the correct one. It is taken literally. But in addition to the two meanings just distinguished there is a third. As in these other examples of metaphor, paralleling the literal sense of the story told is the meaning in action or emotion. The story inculcates attitudes, moral or otherwise, and thus directs action in those who hear it. This pragmatic meaning may continue

2. "Modus per figmentum vera referendi," from Macrobius, *Somnia Scipionis*, I. ii. 7, quoted by Edgar Wind, *Pagan Mysteries in the Renaissance* (Yale University Press, 1958), p. 190 n.

as a lively operative principle long after the story has been exploded as myth. We may continue to perform rituals although we
have ceased seriously to believe. But we act *as if* we believed.

In this brief sketch I have said that usually there is no make-believe in the first stage of the life of myth. But there can be. A great
metaphor made by a genius, and treated as metaphor, always tends
to pass into a later stage in its life. The more effective it is in the
realm of make-believe the more seriously it comes to be believed
in until it is taken literally, by posterity, or by the great man's
contemporaries, or even by himself. It awaits an iconoclast from a
later age who will explode it as myth. Many of Plato's so-called
myths are really allegories or parables invented by him. Some,
like the "Myth" of Er, were intended not for literal consumption
but to function as lively operative principles in our lives and actions. Others, like the "Myth" of the Earth-Born, Plato tried to
foist upon later generations: "This is the Myth. How are we to
get them to believe it? The generation to whom it is first told
cannot possibly believe it; but the next may, and the generations
after. Thus the Public Good may be served, after all, by our Noble
Fiction."[3] It is difficult to decide whether such fictions, neutral in
their nobility, as Force, Gravity, and Attraction were intended
by Newton for literal consumption or to be treated as metaphors.
Although he remarked with approval that the ancients had tacitly
attributed gravity to the immediate action of God, he later attributed forces or powers to the small particles of bodies.[4] In
neither case, unlike Plato, did he give any hint of their make-believe quality. Owing largely to their power as lively operative
principles, unlike Plato's noble fictions, they came to be taken literally in the same generation as well as in the generations after.
Another factor aiding their survival was that they shared with the
most enduring myths of antiquity the metaphor of personification.
At any rate, the myth of bodies attracting and repelling each other
survived long after the myths of Phlogiston and the *Élan Vital*
had been exploded. Their pragmatic meaning still holds.

3. *Republic*, 415 d.
4. *Opticks*, Queries 28, 31.

3. Presenting the Literal Truth

It seems that, accompanying our awareness of the presence of
hidden metaphor, there should be awareness of the literal truth.
A myth is not a fairy story. It has a basis in the facts that the
myth "explains." The *Life-Force* is grounded in the facts of evolu-
tion, the Myth of Reminiscence in all our *déjà vues*, the Theseus
Myth in the wooden bull and the Palace of Knossos, the Myth of
Volitions in all our decisions, and the Myth of Bodily Attraction
in the way bodies move and the way *we* are attracted to other people. To explode a myth, it seems, is not just to become aware of
the metaphors involved but to say what the facts are.

On rare occasions when a metaphor has long been taken literally, a distrustful iconoclast has managed to become aware of the
presence of the disguise. By revealing its presence he has to some
extent exploded a myth. Not content with this destructive achievement, however, he has often tried to do more by presenting the
literal truth. Realizing that the original metaphor involved the
transference of the facts from their proper sorts to improper ones,
he has tried to put Humpty-Dumpty together again by restoring
the facts to the sorts in which they actually belong.

This operation I now illustrate from portions of the arguments
of two philosophers of mind: Plato and Professor Ryle. They first
demolish mistaken theories, and then produce their own. In the
Phaedo Plato refutes two "scientific" theories of the nature of the
soul: (1) the epiphenomenalist view that the soul, being the
harmony of the body or the way the parts of the body are organ-
ized, dies with the body; and (2) the "mechanistic" view that the
soul outwears its bodily garment, but the soul, obeying physical
laws, itself wears out in time. Plato demolishes (1), first, on the
grounds that it is a mere metaphor. Then, by extending the
"harmony" or "organization" metaphor he shows that the additional properties deduced fail to accommodate our ordinary beliefs about the soul. Against his own theory, however, he omits
to use this illuminating technique. In the dialogue itself he finds
no metaphor in his own "substance" doctrine of the soul, the soul
which is permanent, indestructible, and alive, which must be alive

because "a dead soul" is self-contradictory. This doctrine is presented as the literal truth about the soul. Plato's argument against (2), which accords with the theme of this book in its deliberate use of "myth," I shall make use of at a later stage.

A better illustration is Gilbert Ryle's *The Concept of Mind*. In it, Ryle corrects a category-mistake. This mistake is mind-body dualism or "the dogma of the Ghost in the Machine." It is a category-mistake because it "represents the facts of mental life as if they belonged to one logical type or category (or range of types or categories), when they actually belong to another."[5] It might seem that Ryle regards all metaphors as mistakes, for these words just as appropriately define what happens when we make and use a metaphor: they refer to sort-crossing and make-believe, the sort-crossing being that between higher and lower types called "synechdoche." But that he does so is unlikely. Ryle is doubtless aware that it may be most illuminating to account for the facts of mental life in terms of another category. Perhaps this is the only way that we can account for them. Accordingly, the category-mistake that Ryle corrects is probably not just sort-crossing but a peculiar sort of it taken by its victims to have cosmic validity— in other words, what is taken to be *the* correct ontology of the mind. Ryle has thus recognized the presence of a metaphor that has had many victims. In this he is the Wizard of Oz himself, for he is not taken in by the green spectacles that have been worn without awareness by so many since the time of Descartes.

He has therefore "exploded a myth." For him this means more than recognizing the mistake. It means "not to deny the facts but to re-allocate them."[6] And "re-allocation" means more than transferring the facts from one sort to another. It means restoring mistakenly allocated facts to where they "ought to be allocated," [7] where they "actually belong."[8] He thus uncrosses the sorts mistakenly crossed and restores the facts to their proper sorts. What this amounts to is that, while others have been victimized by metaphor, he has not only detected its presence but is able to penetrate

5. *The Concept of Mind*, p. 16.
6. *The Concept of Mind*, p. 8.
7. Ibid.
8. *The Concept of Mind*, p. 16.

the disguise, and to present the literal truth. The latter is a claim to present neither a better theory of the mind nor even a theory. It is a claim to present the correct ontology of the mind.

Ryle's double procedure corresponds to the first two levels of my analysis. At the first level he notices that in the official or mistaken theory such entities as the will, volition, vanity, and the mind are regarded as mental existents, some of which cause effects in the physical realm. Thus the impulse of vanity causes acts of boasting, while the mind itself exists in order to house impulses, volitions, and other incidents. These entities Ryle exposes as myths or occult qualities on a parity with phlogiston, vital force, etc. At the second level, part of Ryle's re-allocation consists in replacing causal explanations by explanations in terms of reasons and incidents; the reasons being expressed as hypothetical or lawlike statements in which such terms as "volition," "vanity," etc., are eliminated because they denote no observables; and the incidents being treated innocuously as cues. Hamlet's remark, "Had he the *motive* and the *cue* for passion that I have, he would drown the stage with tears," characterizes this kind of explanation. Thus instead of occult forces causing occurrences we have the streamlined procedure of premises entailing conclusions. For example, the sentence, "He boasted from Vanity," [9] which suggests that Vanity itself exists and caused his acts of boasting, is interpreted in terms of what might be called "The Law of Vanity," which explains particular occurrences called acts of boasting. This "Law," having been induced from, is testable by, observables. Compare this with my second-level analysis in which the statement "x attracts y" was replaced by: "x and y move according to the Law of Attraction." To choose another example, assertions about a person's mind[10] suggest to Descartes and his followers the existence of a "second theatre of special-status incidents" forever hidden from outside observers. Ryle resolves the problem into "the methodological question, how we establish, and how we apply certain sorts of law-like propositions about the overt and silent behaviour of persons"—"establish," because it is an inductive process arriv-

9. *The Concept of Mind,* pp. 89, 90.
10. *The Concept of Mind,* pp. 167–69.

ing at reasons, and "apply," because the reasons explain or account for behavior.

These examples show how Ryle begins to re-allocate the facts.
But he does not stop here. He provides not merely a new methodology for treating the mind but a new ontology of the mind. For
in addition to providing premises from which conclusions about
the facts are derivable, all of which constitutes an advance upon
the official theory that he corrects, Ryle forthwith proceeds to
identify the referents of his premises with new entities, specifically,
dispositions, propensities, or tendencies. Vanity is identical with
a disposition to boast, etc., and the name "my mind" "signifies my
ability and proneness to do certain sorts of things."[11] The re-
allocating has come full circle. The facts about the mind have been
re-allocated to the sorts in which they actually belong. But just as
Plato did not apply the technique he had used against other theories to his own "substance" view, so Ryle does not turn his own
superb technique against the dispositional account of mind. Dispositions might turn out to be just as ghostly as substances and
other occult forces. All these accounts—Plato's, Ryle's, and their
opponents—claimed by their authors to describe the unmade-up
face of the truth, are really different theories. None of them can
validly claim to provide the correct allocation of the facts. Has
God vouchsafed to any one of them a special revelation? Nevertheless one may paint a more appealing picture than the others
or prove a more useful guide.

The conclusion of this section is a simple one. The attempt to
re-allocate the facts by restoring them to where they "actually belong" is vain. It is like trying to observe the rule "Let us get rid
of the metaphors and replace them by the literal truth." But can
this be done? We might just as well seek to provide what the poet
"actually says." I have said that one condition of the use of metaphor is awareness. More accurately speaking, this means *more*
awareness, for we can never become wholly aware. We cannot say
what reality is, only what it seems like to us, imprisoned in Plato's
cave, because we cannot get outside to look. The consequence is
that we never know exactly what the facts are. We are always
victims of adding some interpretation. We cannot help but al-

11. *The Concept of Mind,* p. 168.

locate, sort, or bundle the facts in some way or another. Thus the second level of analysis which purported to reveal process and procedure purged of metaphysics was at best an approximation.

4. The Geometrical Model

We are now at the final level of analysis where, having become to some degree aware of the metaphors involved, we proceed to re-use them instead of being used by them. But need we use the same ones? If we are aware, we can stop and think. We can choose our metaphors. We are no longer duped citizens of the city-state of Oz but the Wizard of Oz himself.

Perhaps the best way to avoid being victimized by a metaphor worn out by over-use is to show that it is expendable. The best way to do this is to choose a new one. If the operation, just described, of presenting the literal truth is naïve, this one is sophisticated. Perhaps it is better to overleap the last operation, for it is difficult to find the criterion for testing what is reality. Is it revelation? Is it intuition? Is it *esse-percipi?* Or is it ordinary language?

We have seen that the sciences are riddled with metaphors, but the scientists who use them, for example, Descartes and Newton, do not always admit to their use. If they use old metaphors, we can either use theirs or invent new ones. At this level of analysis, accordingly, the search is for the best possible metaphor. We saw earlier not only that there are often rival hypotheses used to explain apparently the same phenomena but that there are tests that can be used for choosing between them. Most of these rival hypotheses involve the use of metaphors, and if these have passed into dogma they involve rival metaphysical views. Accordingly, the choice is between different metaphors. By making this choice, we may be choosing between different metaphysical views.

It is important to make this choice. It is often held, however, either that it is unimportant or that it cannot be done. Some say that a metaphysician adds nothing and takes nothing out. He adds nothing to existence and leaves everything as secure as ever. A piece of physical bread will continue to stay one's stomach just as well as a piece of metaphysical bread whether it consists of occult or hidden forces, mesons or protons, or whether it is merely an idea in the mind. In other words, it is said that a metaphysician offers only a new language for talking about the same things. But we might just as well say that a formal deductive system, like the one lying behind Euclid's system, can be interpreted in only one way. A new theory, even one supposedly not metaphysical, like a new pair of spectacles, changes the facts. It is a mistake to talk of "stubborn irreducible facts." New theories not only save the appearances; they change them, and even create new ones. A new metaphor often lies at the heart of a new theory. That the choice between rival metaphysical theories is possible becomes apparent once we have recognized the metaphors involved. There are criteria for choosing between rival metaphors.

Consider the metaphor, or heap of metaphors, known as "the geometrical model" whose presence we can now detect in the systems of Descartes and Newton. Hallowed in science, it has been so constantly used in various fields that it is now nothing but a disguise placed on the face of nature, a disguise so complete and so ingeniously contrived by a succession of make-up artists from Pythagoras through Euclid to Descartes, Newton, and beyond, that most of us are fooled by it. Newton called it his "Mathematical Way," and Descartes "The Geometrical Method" or *Mathesis Universalis*. But Descartes' is the best model because, of all the make-up artists, he was most aware of what he was doing. On the night of November 10, 1619, having experienced a moment of illumination, Descartes dreamed a dream which, after interpretation, enabled him to envisage the extension of the geometrical model to every subject. All his subsequent work amounted to presenting outlines of, and exercises in the use of, this model. In what follows I abstract features needed for my purposes.

The first "geometrical" property specified in the geometrical model is deduction, the same as Euclid used, but this property is to be used beyond abstract geometry to apply to everything: "Those long chains of reasoning, all simple and easy, which geometers are wont to use to reach the conclusions of their most difficult demonstrations, had led me to fancy that everything that can fall under human knowledge forms a similar sequence."[12]

The second "geometrical" property specified in the geometrical

12. *Discourse on Method*, II.

model is extension, but, amazingly, this property is to be treated as the defining property of the physical world, for the latter *is* nothing but *res extensa*. Physics is geometry, or, in other words, the object of abstract geometry, namely extension in length, breadth, and depth, is the object of physics: "But I have only decided to give up abstract geometry . . . in order to have more time to study another kind of geometry which concerns itself with problems about the phenomena of nature. . . . My physics are nothing but geometry."[13] But so far this physics is merely statics without force or even motion.

The third property, therefore, is motion, which is a mode of extension. Although this is not an object of "abstract geometry," it is specified in the model. Its inclusion, it seems, enables the first three properties to define mechanics, making the geometrical model identical with the machine model: "I have hitherto described this earth, and generally the whole visible world, *as if* it were merely a machine in which there was nothing at all to consider except the *shapes* and *motions* of its parts."[14]

The last passage suggests that Descartes' machines needed no force to move them, that is, his mechanics was merely kinematics. The *kinema* lacks *dynamis*. But we have already seen that he assumed corporeal forces in his laws of nature and throughout his physics. Accordingly, corporeal force must be specified as the fourth property in the model.

The geometrical model thus built is identical with the machine model; and so Descartes conceived it. He was clearly aware of the great explanatory power of his model, and of the power over nature it could give man. His Archimedean rhetoric, "Give me extension and motion and I will construct the world,"[15] abbreviates the properties of his model just specified, with the exception that incorporeal force replaces the corporeal. Imbued with geometrical zeal, Descartes actually did apply his model to solve problems in optics, magnetism, geogony, biology, physiology, and psychology. For example, the human body became "an earthly machine," while emotions, such as love, were nothing but the effects of "some

13. Letter to Mersenne (1638).
14. *Principles*, IV. 118, my italics.
15. *The World*, ch. VI.

movement of the animal spirits." But he stopped short at the mind or soul. He decided that the model was not to be applied here, though it has been since.

From this account we can see clearly how metaphor is present here. The whole method involves the representation of the facts of all the sciences in the idioms appropriate to geometry. Moreover, it involves make-believe throughout; for example, that forces reside in bodies, that the properties of the world are the properties of geometry, and so on. These satisfy the test of metaphor. We see just as clearly that it is an extended metaphor which, because the properties of this metaphor are carefully specified, is a model. Descartes, out of his genius, had invented one of the great metaphors of mankind. It is as though, waking from his dream, he uttered: "Geometry, be my metaphor!"

But I have spoken as though Descartes was fully aware of the presence of metaphor throughout. It is certain that he was partly aware. The passages quoted show it. By specifying in detail the properties that he intended to use, his extended metaphor became what I have called a model. He had adopted or devised, in Black's words, "an extended system of implications." His metaphor was "made-to-measure." It was not "a reach-me-down." Moreover, this metaphor meshed beautifully with the rest of Descartes' system. The only properties that he admitted into his premises were those he could "clearly and distinctly conceive." But it is equally certain that he was in part a victim of his own metaphor. As we saw earlier, like Newton, he was victimized by the metaphor involved in "corporeal forces." Like Newton, he was so intrigued by the deductive chain of reasoning in his procedure that he exported the necessary connection invented between premise and conclusion to the realm of process in nature. He thought that nature obeyed the deductive relation. He would have seen no absurdity in the remark: "Corporeal forces necessarily entail their effects," or the remark: "Principles cause their conclusions," because there would have been no confusion of different sorts. Again, he did not merely make believe that extension in length, breadth, and depth constituted the external world. He took this metaphor literally, for he defined the world as *res extensa*. Finally, he did not merely make believe that human bodies and dogs' bodies were machines, they were literally machines.

In short, enthralled by his own metaphor, he mistook the mask for the face, and consequently bequeathed to posterity more than a world-view. He bequeathed a world.

5. *The Linguistic Model*

But the model he chose, and the model posterity adopted and proceeded to confuse further with the thing modeled, might have been different if Descartes had dreamed a different dream, and had used his genius to interpret it as well as he interpreted his dream of universal geometry. He might have chosen language as his model. In which case, instead of writing, late in life, the passage about the world-machine, he might have written: "I have hitherto described the earth, and generally the whole visible world, *as if* it were merely a language in which there was nothing at all to consider except signs, things signified, and certain rules of grammar."

Had he done this, and had Newton continued to develop his great teacher's insight, we should now be living in a different world. As it is, and as I have already noted, these two great sort-crossers of our modern epoch have so imposed their allocation of the facts upon us that it has entered the coenesthesis of the entire Western world. Together they have founded a church more powerful than that founded by Peter and Paul, whose dogmas are now so firmly established that anyone who tries to re-allocate the facts is guilty of more than heresy; he is opposing scientific truth. To use Lord Russell's phrase, by beating his head against the brick wall of science, he is headed for disaster. For the accepted allocation is now a defining feature of science.

But that it can be dispensed with I shall now try to show. The best way to do this is to show that it is only a metaphor. The best way to show this is to make another metaphor. I shall therefore treat the events in nature *as if* they compose a language in the belief that the world may be illustrated just as well, if not better, by making believe that it is a universal language instead of a giant clockwork; specifically, by using the metalanguage of ordinary language consisting of "signs," "things signified," "rules of grammar," and so on, instead of the vocabulary of the machine consisting of "parts," "effects," "causes," "laws of operation," and so on, to describe it. Now that we know we are choosing between different metaphors, tests rarely resorted to reveal themselves. But it would be premature to apply them without first putting the metaphors to work within a specific subject matter where they can be used to produce different theories capable of being tested by those canons customarily used to test any scientific theory.

Accordingly, having clarified the foundations of my method, and having chosen my new metaphor, let me take a concrete problem that has bothered and intrigued scientists and philosophers from the time of Plato, the problem of vision, in order to see how successfully they can be applied. But it will be necessary first to elaborate the main features of the linguistic metaphor.

PART TWO

THE METAPHOR DESCRIBED AND APPLIED

CHAPTER IV

Ordinary Language

1. Signs

ALL OUR KNOWLEDGE has to do with signs. We notice things, but noticing is not knowing, and what is known is not noticed. "We know a thing when we understand it, and we understand it when we can interpret or tell what it signifies." Macbeth did not *know that* the three witches were women because their beards forbade him to *interpret* that they were so. Signs invented by humans may be called artificial, the rest natural. Natural objects, however, are not signs, although they may become so. Similarly, spoken words are signs, but uttered sounds may become words. However, if there were no perceivers, there would be no signs, not even natural ones. The existence of signs consists, therefore, in being perceived and interpreted. What then, are the interpretations of objects, or what do signs signify? Often they suggest absent things to us. Certain clouds suggest rain. But this is an abbreviation. For, although we think of rain, the rain itself is absent. Present to our thoughts beside the clouds we see there may be images. More often than not, however, objects become signs without suggesting other things at all, and without any images coming between. They can become cues for passion or for action. Being directed how to act or to get ready to act by means of signs is interpretation of them just as well as being minded of other things by their means. It is knowledge. It is knowing how to act or to prepare for action (*S*, 253).

We are prone to overlook signs, for our interest is centered in what they signify. This causes telescoping in our thoughts and abbreviations in our words. In reading we overlook the letters for their meaning. We say that we see words and even propositions and meanings, although we see only marks which suggest these to us. We say that we see gladness in his looks and shame on his face, that ice looks cold and water wet, although we know that these emotions and qualities are colorless. In cases such as these, the signs and the things signified are so closely blended or concreted together by frequent use and the immediacy of the connection that it is difficult to separate them in our thoughts.

Men invented artificial signs or symbols out of convenience to assist their thought. In most cases the sign seems to be more easily comprehended than the thing it signifies, which is often difficult to comprehend. We use things we are directly acquainted with to stand for things we know by description, things present to the senses for things only imaginable, sounds for thoughts, written letters for sounds, small things for big things, concrete things for the abstract or spiritual, and so on. To assist our thinking, we make allegories, parables, and metaphors; models, diagrams, and gestures; flags, emblems, counters, and coins. Plato used the state to represent man, and he used a good horse, a bad horse, and a charioteer to stand for the will, the appetite, and the reason, respectively. Freud used other devices to stand for the mind. Descartes used the pilot of a ship, while Locke used an empty room. Logicians use symbols to stand for thoughts that would otherwise engage the higher faculties of the brain (*A*, 7).

Words are artificial signs. Accordingly, being invented for "quickness and dispatch" to illustrate other things, words and their referents are, in general, quite different in kind or heterogeneous. Signs of things do not have to be of the same kind as the things they signify. Moreover, in general, words are not pictures or copies. They are not necessarily icons that mirror the world as a footprint does a foot or Caesar's bust does Caesar. Words do not need to be pictures although, like "chirrup," "gurgle," and "guttural," they can be. Nor are they necessarily connected with their referents. Other words might have been used instead.

There are, however, some difficulties in these views. I have said that a word is in general quite different in kind from what it stands for, and yet, in speaking, we call both by the same name. When things are called by the same name they are often thought to be the same thing or of the same kind. The dog I owned once and the dog I own now I call by the same name. They are the same dog. The particular dog, Yackamundi, has another name too, which he shares with others, namely, "dog," while all dogs are homogeneous in that they belong to the kind called "dog." Since words and what they stand for are called by the same names, perhaps they too are homogeneous. However, when things are said to be of the same kind, they often have some of the same properties or are alike in some respect. I have denied the picture theory of meaning. But it is certain that by far the greater number of words are either the same as or very like what they name. The word "dog"names dog; " 'dog' " names "dog"; and so on. And the same noise names not only what a noise stands for but the noise itself and its name too.

2. How Different Things Get the Same Names

Many different things with the same names are said to be the same things or of the same kind, sort, or class. Every glimpse of the dog Yackamundi gets the same uncommon name; every kelpie gets the same common name; and every kelpie, dachshund, and dalmatian gets the same very common name. These things are thought to be the same not only in name. We say that they belong to the same kind and that some of their properties are the same. But they also have many differences, and each one is a different thing. Moreover, we do not usually think that they get their homogeneity and their common properties from their common names. So the question arises: How do many different things get the same name?

The first common way is this. Different things get the same name such as "dog" or "same sort" because some of their properties are the same, and this, in turn, because the different things are more or less alike. While things may get the same name for these reasons we cannot be certain of the reasons when we know the name. For some things with the same name are not of the same sort although they have some of the same properties and are alike; some have only a superficial likeness; some are the same or alike only in the name; and some only in the sound of the name. Indeed, few words are univocal. "Winston Churchill" names the man in Chartwell and another in Rochester, but in the catalog they go for men or are said to belong to the same sort.

The second common way is very like the first. Some things—very different this time—get the same name because, although they have very few things in common, either they may have a special likeness that someone is interested in or someone is interested in creating this likeness. But they are usually not said to belong to the same sort. Often the same properties may be thought to be only one. When this occurs, the common name that results is often a metaphor. Metaphors, freshly made, are a deviation from common usage, the old name newly shared, being, according to the *Shorter Oxford English Dictionary*, "not properly applicable" to the new object.

In dead metaphors, such as "perceive," "comprehend," and "metaphor," however, the questions of homogeneity and likeness do not arise because, although the etymon overlaps its metaphorical meaning, these are overlooked by all but scholars. In dormant metaphors also, such as "high note," "to see meanings," and "to lay bare feelings," both meanings have become literal. There is some likeness between a high note and a high wall, but their common heights are not usually thought to be of the same kind. In live metaphors both meanings are present to thought. Although the metaphor "metal fatigue" illustrates a characteristic of metals, the wearinesses of muscles and metals are thought to be heterogeneous. It is the same with other metaphors such as "in the mind" and "on the nose." Thus metaphors make a class of cases in which things with the same name are thought to be different in kind. Nobody thinks that Mussolini was one of the many members of the class of utensils although, for many people, he now shares their name.

Different things, as already noted, often get the same name through their possession of the same properties or through resemblance in some respect. One way to make a metaphor is to be the first to notice these and then to disguise a mere comparison by giving the different things one and the same name. This involves using an old name to name a new object, or, which is the same thing, adding a new referent to an old name, thereby illustrating the new object by drawing attention to the common property or properties owned by the referents old and new. He who penetrates the disguise calls the name thus used a metaphor. He who does not takes the metaphor literally, for he takes the disguise for the face of the truth. If this occurs, he adds additional properties of the sign to the thing signified. Unlike making a name by going from the properties to the name, it involves arguing from the name to properties.

There is a third way in which different things get the same name. Things may get the same name merely through their co-existence and convenience. This is the case with many signs and the things they signify. Other properties besides co-existence may be shared but not necessarily. In a wide sense the resulting shared name is also a metaphor because at first, the name is "not properly applicable" to the new referent. In metaphor, however, it was noticed that at least one property, the name, of the things literally signified was transferred to the thing illustrated. If the metaphor was taken literally, additional properties were transferred. But now the case is reversed. The name of the thing signified is conferred on the sign. As in metaphor, if the old name newly conferred is taken literally, other properties of the thing signified as well as the name are conferred on the sign. We act as though the things signified, not just the signs, were present to us. For we consecrate, hallow, and display, not only the spirit of the regiment and the spirit of the departed but also flags, altars, and ground. We revere not only the spirit of England but this blessed plot. We see on her face not only a look of gladness and a blush of shame but also gladness and shame. We hear not only sounds but words, propositions, lectures, bells, and aircraft. We see not only colored shapes but words, meanings, and the points of jokes. Ice not only feels cold and water wet; they do indeed look cold and wet. There is not much likeness between a blessed plot and a blessed spirit, between a blush and shame, between sounds and a proposition, between marks and a word, between a colored shape and a joke, or between a look and a feel, except in name. Although we speak in the way of the examples, we do not take our words literally, for we know that language itself is framed "for ease and expedition," and that, out of convenience or laziness, we have abbreviated further. Why invent a new name for the sign when we already have one for the thing signified where all our interest lies? It would be redundant and contrary to the design of language. And we are usually understood.

Here then is yet another class of cases in which very different things get the same name. Some people, however, may not accept some of my examples. They may think that ice does in fact look cold, and thus that looks and feels are homogeneous. They may also think that we see and hear words just as clearly as we see and hear the marks and sounds that name them. These pairs are indeed most closely twisted, blended, and incorporated together, and this "complication" of each pair is reinforced by their constant association.

Words and their referents, it seems, are of this sort. We do not think that they are homogeneous although a word and the thing it names are called by the same name. To things that go together very frequently we tend to give the same name, for example, to many signs and the things they signify when these are closely blended or "complicated" together. To a word-sign and its referent we do so always, at least if we do not regard quotation marks as part of the word. But as before, although they have the same name, we think of them as different things. A stranger to kelpies and Siamese cats but not to cats and dogs can recognize either as a dog or a cat as soon as he sees one. If the marks DOGS and dogs were closely alike, a stranger to the English tongue would be able to tell what the mark DOG or its corresponding noise meant as soon as he saw or heard them, just as a foreigner thinks of *man* as soon as he sees the marks the Chinese use to mean it. If the name of dog were a kelpie or ◁▯◻▯ , he would have some idea of what it meant, just as we can make sense of some hieroglyphics. It might be asked: Why do DOGS and dogs get the same name if they are so dissimilar? The well-known answer is that this is a convention based upon convenience and economy, for it saves an endless number and confusion of new names. Since our main interest lies in what is marked by the name rather than in the name, it would have been superfluous to have given the sign a distinct name. Although there is involved the risk of confusing heterogeneous natures, that is, of category errors such as deducing homogeneity of nature from homogeneity of name, this convenience has become conventional.

We often speak *as if* there are "natural" sorts or kinds. This way of speaking is, to a large extent, unavoidable; for language reflects our convenient errors. Nobody thinks, except the New Guinea tribes who make raids to capture them from others who own them and wear them at their belts, that we find names in the world. But in a sense, things come with their names on them. We "find" words already standing for things. We also "find" recurrent characteristics, connotations, things already numbered and sorted out, and necessary connections already threaded for us. They are there *a priori,* but not prior to all experience, merely prior to ours. These categories are, as it were, built into our minds through a long tract of time from the moment we appear. They are part of our cultural, not our natural, heritage.

We all agree that we find in the world particular things, some co-existing, some succeeding others, and some more or less alike. But it is almost impossible to separate, or even to distinguish, what we actually find from what we "find," for the raw stuff is overlaid with the contributions of men. And these contributions have come largely through the super-impositions of ordinary language. Particulars, having been observed merely to go together, have already been combined by our forebears and the result bequeathed to us as one distinct thing signified by one name. The numbering names "one" and "same" are already affixed, so that now we cannot easily see the thousand leaves for the one tree. We are enthralled by the fixed and steady names of things and by the ways in which they have been combined, named, and numbered. Many particulars, having been observed to be closely alike, have also been collected and combined by our practical ancestors, and the result has been bequeathed to us as one distinct sort or kind signified by one name such as "of the same sort" or "dog." Any very close resemblance was, for "convenience and dispatch," subtracted from the particulars and called "recurrent characteristic," or "common property," or "the same thing" happening over and over again.

All this numbering, naming, and combining together of things and qualities was perfectly arbitrary. But the results and the manners of getting them, that may have been, at first, queer or unusual, then fashionable, then customary or conventional, became our inheritance. We are imposed upon by words. We find, already laid out for us, not only the conventional meanings but the numbers, the recurrences, and the sorts. It is unlikely, however, that a Martian would parcel out the things of this world into the same distinct collections that we do.

Naming, numbering, or sorting things is not just noticing what is out there fixed and settled. It is the result of looking through conventional blinkers. Nevertheless, there are arguments about sorting. These are mainly verbal. There are few about tigers and lions. There may be some about "tigers" and "lions." We do not remain in disagreement for long about the marks of this tiger- and that lion-like animal. Is it a sort of tiger or a sort of lion? Or is it a new sort? The convenient way chosen for the tigron was the last. We can make new sorts as we please. But those that we have grown accustomed to, we tend to think are determined and set out by nature. These also were grouped and named in an arbitrary manner. They might have been sorted in a different way.

3. Definitions

Most definitions also, are part of our inheritance. They help us to understand, by means of words, the ways in which a preferred class of men have used words to stand for other things. There are many sorts of definitions. Two common ones are the denoting and the connoting. The first delimits a sort by providing for us some of the smaller sorts included in it, or by providing some of the particulars that come under it. For the genus *dog,* the species can be given, or the breeds, such as kelpie and dalmatian, etc., or the particular dogs such as Melampus, Yackamundi, etc. And we are implicitly instructed that any one of these sorts or particulars belongs to the sort *dog* as well. The second delimits a sort by providing some of the larger sorts to which it belongs, and then by abstracting their common properties. But, more usually, as is the case in most dictionaries, this is left to us. The sort *dog* is included in the sorts *animal, quadruped,* and *barker,* or in the sorts *carnivore, non-climber, eyes in front of head,* etc. Or the sort *triangle* is the common property of the sorts *surface, plane, three, closed,* and *straight lines.*

This second way, because it omits details and does not specify particulars, seems to describe new entities intervening between the common names and their many particular referents. We thus tend to suppose a single, abstracted, precise signification described by the right-hand side of each definition and annexed to every common name, such as animality or four-leggedness, or doggishness. This precise naked entity has been given various names: "defining property," "connotation," "designation," "abstract idea," "Idea," or "universal." Some suppose it to be manufactured by the understanding and to exist only in the mind; others that it resides outside the mind and only in particular things like the redness found in a tomato or a cheek; while others suppose that it exists both apart from things and outside the mind. But all who subscribe to the doctrine agree that this entity is what we use to recognize instances of it because they manifest it or partake of it. This strange doctrine persists in one form or another among intelligent people although, try as one likes, one cannot find in sense, intellect, or imagination, the abstract entity in question.

It seems, however, that there is no need to look for it, for we do not need it. We can manage perfectly well with words or names, their definitions (which are nothing but sets of alternative names), and the particulars that some words stand for. For it is one thing to keep a name constantly to the same definition, and another to make it stand everywhere for the same, precise, discrete entity, whether it be an abstract idea in the mind or an abstract universal outside the mind. The one is necessary, the other useless and impracticable. "And he that knows that words do not always stand for things will spare himself the labor of looking for them when there are none to be had" (*I*).

We can describe the denoting way of defining without recourse to sorts by saying that the name "dog" is defined by giving a list of less general names, such as "kelpie," "dalmatian," etc., or such as "Melampus," "Yackamundi," etc. In which case we are implicitly instructed by the definition that any of the things thus named "has the right to" the name "dog" as well. As a result, we learn, for example, that Melampus has at least two names, or we learn that the particulars named by "dog" have other names as well. The connoting way may be described without recourse to sorts or abstract entities by saying that anything named by "dog" has the right to be called by any of the names on the right-hand side of the definition as well, that is, by the name "animal," "four-legged," etc. Although the right-hand side may appear to name an abstracted entity that no one can find, this is just another abbreviation which must not be allowed to become "a cheat on the understanding." It is just that the definition leaves out a lot. It is not said whether the dog is big or small, dalmatian or kelpie, spotted or brown. The definition merely instructs me that whatever particular dog I consider, the sentence "He is a four-legged animal that barks" holds true or that he has the right to a host of other names (*I*).

But it may be asked: How does it happen that particular things get their common names unless they have a common essence or property and belong to pre-determined sorts or kinds? Part of the answer has been suggested. For I have already sketched an account of some ways in which different things get the same name. This is merely the reverse way of sketching an account of how words become general. Particulars (including marks and noises) become general probably from a host of different factors. Some of them are these: A particular becomes general by being made the sign of another particular, and this because they constantly and closely go together. Again, a particular becomes general by being made the sign of many other particulars between which there is some likeness. (There are, of course, difficulties in this "resemblance theory of universals" which are still a matter of controversy, but it seems likely that these are soluble without recourse to abstract entities. This is a problem, however, with many complexities, which it is not the purpose of this book to unfold.) Another factor influencing us to parcel and bundle particulars into sorts in the way we do, must be, as I have indicated, that we are heirs to many ages. We tend to leave tied the parcels and bundles previously made by our forebears. The factors that influenced our forebears might have been no better or worse than whims or fancies. We could easily influence our children to bundle Siamese cats and lions together with dogs instead of with cats, and kangaroos with wallaroos. But when things become signs, that is, when particulars become general, as was previously noticed, we pass from the realm of merely noticing into knowledge. A foreigner to a language hearing it spoken for the first time cannot be said (except in an abbreviated way) to hear words but only noises. For these noises, being entirely terminated within themselves, have not yet become signs (*I*).

4. Necessary Connection

We are prone to think that signs have in their own nature a fitness to represent as they do. Between words and the things they stand for the connection is imposed by us, but we often forget it. It is difficult to hear the common words of our native tongue without understanding them. It is likely that, as Voltaire said, if all men spoke the same language and were born with the faculty of speaking it, we should always be inclined to believe that there is a necessary connection between words and things. The choice of light and sound for our artificial signs was arbitrary. Smell and taste might have been used instead. It was also arbitrary what particular combination of sounds should have been used to stand in place of absent things, and that marks should have been used at all to represent these sounds. The words of any language are, in their own nature, indifferent to signify this or that thing at all. We can find nothing in a word itself which enables us to bridge the gap between it and what it stands for. But when the use of a word has become conventional, its subsequent application is not arbitrary.

Men have given the name "necessary connection" to a certain relation between artificial signs. Like sorts, names, and definitions, this relation has been put there. Consider two cases: the marks DOG and a particular musical score—so brief that it can be read in an instant—which represent respectively the sounds dɔg and a musical performance. The former seem peculiarly appropriate to represent the latter. Both members of each corresponding pair not only have the same name; they have the same number of parts. Moreover, they *must* be as they are or, at least, very closely like it. But this necessity is the end result of: first, the arbitrary selection of certain particular sounds and of certain particular marks to represent them; secondly, their customary use and acceptance as conventions; and, finally, the setting down of combinations of marks to fit the sound combinations arbitrarily chosen. Thus anybody who now uses different marks in any cases such as these either refuses to abide by the conventions or accepts them and contradicts himself. In other words, if anybody knows the written and the phonetic alphabets in such cases, knows how to use them, and accepts them, then he must use the appropriate marks for the new words. These marks, however, considered in themselves or in their own nature, have no necessary connection with anything.

The statement "cats are feline" is conventionally called a necessary truth. This is because the lexical definition, " 'cat' means 'feline,' " is a conventional expression of the custom of calling the same things by the interchangeable names "cat" and "feline." Thus anybody who now says that some cats are not feline is either unconventional or he is conventional and contradicts himself. If, on the other hand, any non-conformist defines "cat" to mean "canine," and sticks to it, then, if he says that cats are canine, he utters a necessary truth.

In order to show how necessary truths are contingent upon previous decisions and agreements I shall perform two experiments.

In the first, I suppose a most intelligent although highly conventional illiterate who is an expert in phonetics and music. He has already learned and accepted the conventions about naming, numbering, and sorting, and, being an expert, he knows perfectly well how to discriminate sounds and to number them. Thus the sounds dɔg, which, perceived at the same instant, are apt for many of us to coalesce into one sound, for him are discernible as three distinct sounds. Now it might be argued that there are not three sounds here but only one; that this sequence is, therefore, nothing but a monosyllable. Certainly such a sequence is traditionally called a monosyllable. But a monosyllable is often a combination of many sounds slurred together just as a monogram is a combination of many letters interwoven. I adopt here the view expressed by Socrates in the *Theaetetus* (204) that "the syllable is a simple form arising out of the several combinations of harmonious elements." Let us now suppose this expert confronted for the first time with the marks DOG or the musical score mentioned earlier. For convenience in presentation, I shall treat only the former. It is obvious that he would be bantered if asked to find any connection between the marks and the sounds he knows so well, his view of the marks being entirely terminated within themselves. And if, from this moment on, he were to be left alone, he would never find any. But after the experience of learning from others to put together the written and phonetic alphabets, he would come to regard the marks DOG as a sign for the sounds dɔg and would follow the convention in giving them the same name. Nor could he help giving the marks the same number as the sounds. And he would, in time, regard the marks as most fitting to represent the sounds. Then he would realize, perhaps with astonishment, that the numbers of the marks and the marks themselves *must* be as they are once he has accepted the second alphabet. But he would also realize that these marks and sounds were of human institution, that there was a time when the marks were not connected in his mind with the sounds they now so readily suggest, but that their signification was learned by the slow steps of experience, and that, accordingly, they might have been otherwise. Thus he would be preserved from confounding DOG with dɔg. They would never coalesce into one distinct thing—as dɔg do for us—although they have the same name, the same number of parts, are necessarily connected, belong to the same sort, and suggest the same things.

The second experiment is the same as the first except that I suppose these marks and sounds to exist in nature, each pair always going together and, instead of the illiterate, I suppose a blind man made to see for the first time. The other ingredients are the same. As before, it is arranged that the sounds are out of earshot immediately after the blind man is couched. The events unfold much after the same manner as before. At first sight he would be bantered if asked to find any connection between the marks and the sounds he knows so well, for he would be unable to isolate—let alone find—the marks. Then, the sounds having been brought within earshot, after protracted experience, he notices no connection between the two except that they go together, his view of the marks being entirely terminated within themselves; nothing he sees being able to suggest to his thoughts the ideas of high or low, loud or soft, near or far, sharp or clear. Later, the constant and long association of DOG with dɔg in sense and imagination makes him judge the former as a sign of the latter. But he is already being taught by others that they are not two things but one and the same object with one and the same name having two sets of sensory qualities, one visual, the other auditory. He is also taught that (it is proper to say that) we are able not only to hear the sounds but to see them; not only to see the marks but to hear them; in short, that we see and hear the same thing. Being conventional, the once-blind man accepts all this. But it is doubtful whether he notices that it is conventional or how convenient it is. He regards the marks and sounds as naturally fitted to go together, and, on being asked why, says that, for one thing, they have the same number of parts. He can see that the object has three distinct marks answering to the three distinct sounds it makes when he listens. He thinks that the marks *must* have the number they do have, not because of any prior agreement but because he can see them; and he would be astonished at the suggestion—unless he was given to reflection—that they might have been numbered otherwise. Moreover, he now thinks that the marks themselves *must* be as they are and could not be otherwise. Finally, he might remember that in the past he thought that DOG and dɔg were entirely distinct and heterogeneous. But if so, he now thinks this was a mistake.

It was noticed that the once illiterate contradicts himself if he calls the sounds dɔg by any other written name than DOG once he has learned and accepted both alphabets. He is no non-conformist and he is self-consistent. It must be acknowledged that the two words have the same number of parts. Indeed, they must have, for this had been agreed upon before his teachers began their task. To notice that they have the same number just by looking and listening was beyond the illiterate's ability. But this necessity is compatible with the arbitrariness of words.

Although it might seem otherwise, similar remarks apply to the case of the once-blind man. He contradicts himself if he says that the marks are not three in number. His teachers had accepted the well-known conventions about naming, numbering, and sorting. It had been agreed before *they* were born that these sounds and

marks were one and the same thing having three parts both visible
and audible. These parts had been called letters and the whole
object one word. The once-blind man, as we have noticed, accepted as much of all this as he could understand before he was
couched. And he could recognize the object by its sounds. After
being couched, his learning to see (or read) was accelerated by his
teachers. Finding the marks going together with the sounds, he at
first regarded the one as a sign for the other, but he was instructed
to fuse them into one thing. He learned that the one object was
visible as well as audible. And eventually he became adept at
recognizing the object by seeing it as well as by hearing it. So now
he cannot help but see three of the letters every time he looks at
the object, because he has been conditioned to see three. This is a
psychological event. On the other hand, he *must* say that the
marks or letters are three in number. For this had been agreed
upon before he was born, and he is neither non-conformist nor
inconsistent. But this necessity, based as it is on previous agreements and decisions, is compatible with the arbitrariness of the
marks themselves. Other marks than DOG might have gone together with dog. It happens to be the case that these do.

To sum up this account, there are two well-known types of
connection between things. The first is contingent or customary.
For example, it happens that my table is square. In order to find
out whether it is, I must know the meaning of "square," and I
must measure the table. The second connection is necessary. Such
for example, is the connection between my square table and its
four sides. In order to find out whether it is, I need know only
the meanings of "square" and "four-sided" and an elementary skill
or "know-how." No further experience is required. That is why
the truth just found is called *a priori*. I know it with certainty, and
I contradict myself if I say that my square table has only three
sides. The same connection holds between the premises and conclusions of valid deductive arguments: between Euclid's postulates and his theorems, between the constitution of the United
States and its corollaries, or between the words of the Scriptures
and their entailed consequences. If we put together as one statement the subjects and predicates or the premises and conclusions of
any one of these examples, we produce a necessary truth or an

analytic statement or a tautology. Sometimes these names are applied to the conclusions themselves, for example, to the conclusion that a triangle's angles add up to 180°, but to do this is to abbreviate and hence often to confuse people.

In all this we must not be cheated into thinking that necessary connections hold between things considered in themselves, for necessary connections are contingent upon other things, specifically upon implicit agreements, decisions, and previous experience. Certainly I utter a necessary truth if I say that my square table is four-sided. But this necessity is contingent upon my being consistent with my early acceptance of the usual meanings of these words which I had to learn. The connection between all signs and what they signify is an habitual connection that experience has made us to observe between them. And the sounds or marks that men have made into signs, and to which they have given meanings which by custom still obtain, are, in their own nature, indifferent to mean or signify this or that.

5. Understanding Words

This brings us to the question of how we understand words. We have seen that we understand signs if we can tell what they signify. This applies to words. But I must fill in the details. I said earlier that words and what we use them to stand for are heterogeneous, not at all alike, and not necessarily connected. But then we found that some words and their referents are of the same sort, are closely alike, and are necessarily connected. On analysis, however, we found that sorts, the likenesses between words, and necessary connections have been imposed by men out of convenience. So if we consider words in their own nature, that is, before they have become signs, as merely marks and sounds, my earlier remark still holds—at least for heterogeneity and necessary connection. This still leaves likeness between a great number of words and their referents. Although many words are like their referents, nevertheless they make cases in which we are not greatly interested. If we were, we should probably have made the names of names much less like them than we have. What ordinary people are mainly interested in (and ordinary languages were made by ordinary people) are the cases in which words refer to things and not to other words. And in these there is likeness between the two only in exceptional cases. But even if we were to allow that there are natural sorts, we should not even put words and their referents into the same sort. That is to say, we do not say that they are of the same sort. Having made these modifications I think that I may still maintain that, in general, words do not suggest the things signified by any likeness or identity of nature, but only by an habitual connection that experience has made us to observe between them.

It is, then, by experience that we understand words. This truth hardly needs to be insisted upon, for nobody thinks that the connection is innate. Since it is an habitual one, it is learned like any other habit. We learn it as we learn to see (except that with words we are more aware of the learning process) or as we learn to play cricket or tennis. We learn to understand many words at our mother's knee when objects in our presence are named for us. Associations are built up in such a way that when the objects are remote from us in time or space the presence of the words suggests them to us. But what this often amounts to is that, although we say that the words suggest absent objects to us, the words call before our minds images which we believe to be *of* the objects. I can understand the words "The trees are in the park" perfectly well because most of them were so often ostensively defined for me during the first three or four years of my life. These words make me think of the trees in the park. What I do often is to frame images which, so I used to think, were *of* these things, and to which I gave a secret reference to the real things, thinking that, in some way, my mind was able to make this enormous leap. All these words that have been ostensively defined can be "cashed" either in the things themselves or in images; and understanding them consists in being able to do either of these things. On the other hand, although these words are thus "cashable," it is probably most often the case that we rarely "cash" them. For words learned in this way are like counters at a card table. They are not usually cashed during the game although they can be at any time.

There are, however, many words which we learn to understand although they are "cashable" neither in things nor in images. They include the connections like "and," "but," "although," and so on.

But they also include abstract nouns like "freedom," "time," "Grace," "The Trinity," and algebraical symbols like "$\sqrt{-1}$"; all of which some of us, and some of which all of us, have learned to understand. How do we learn them? We learn them through noticing how they fit into their contexts. Thus the word "time," taken by itself, is difficult to understand, but sentences containing it can be easily understood. It is difficult to give meaning to "freedom" abstracted from other words just as it is difficult to conceive freedom, the abstract idea, but we all know, in the gross and concrete, that we are free agents. In reading a page or in listening to the spoken word we do not stick in this or that phrase but collect the meaning from the whole sum and tenor of the discourse.

There is a third way of learning to understand words which constitutes a new dimension. Although we may learn part of their meanings by ostensive definitions or by noticing them in their contexts, we learn another part by responding to them by our feelings or our actions. We could say that their meanings are placed in the will and affections rather than in the understanding, but it would be more accurate to say that the way we respond is part of our understanding of them. The sound of approaching aircraft sent many of us diving for the slit-trench and filled us with fear at the same time but without any images coming between the sound and these two responses. Similarly the words "Six Bettys approaching from the northeast" got the same responses without any time for thinking of their referents. Sometimes the words are understood and their meanings learned merely by feeling about them or by doing something about them. Words like "Faith," "The Trinity," "Grace," and symbols like "$\sqrt{-1}$", "$\supset$", and "$\vee$", we did not learn to understand ostensively. We may have got part of their meaning by noticing how they were used. But many people have come to understand them as instruments to induce emotion and to direct our practice. We learn how to use them in order to regulate our movements and actions $(A, 7)$.

The well-known distinction between *knowing that* and *knowing how* fits the learning of a language. This corresponds to the two ways of learning a language either by rule or by practice. A man may be skilled in language without understanding the grammar of it and without knowing the dictionary definitions. The latter are subservient to the first. We rarely pause to consider them but use them only as checks on our skills. They are distilled out of our practice. Knowing the theory of anything is contrasted with "know-how" in all the arts such as cricket, music, baseball, walking, writing, and seeing. Babe Ruth, Sir Donald Bradman, Beethoven, Greta Garbo, Michelangelo, Don Juan, and Shakespeare, all great exponents of "know-how," probably knew how to manipulate their instruments to achieve the desired results long before they knew the theory of their art. Perhaps some of them never bothered to learn the theory. On the other hand, there are many who know the theory better than these, but who lack "know-how." Many myopes and presbyopes are theorists in seeing. But we do not think that when they see an approaching automobile they do it in virtue of their knowledge of theory. For brutes and children who sucked in the "know-how" with their milk can see better than they can. Children can speak an ordinary language and understand the words of it as well as the best grammarians and lexicographers. There are, of course, notable exceptions. That great theorist in vision, Helmholtz, so applied his discoveries in theory that he was able to perform tricks in seeing that are beyond most of us. Some interpreters of aerial photographs who know the theory of the stereoscope are able to discard it in their work, achieving the same effect with the naked eye.

Although we acquire the skill of understanding words by experience, so that we know the correlations between them and things, between words and other words, and between words and feelings and actions, we do not do it by inductive reasoning. Nor must we think that we do it by deductive reasoning. What may be a vaticination is not necessarily a ratiocination, although it can be. In the main, words are cues rather than clues. An actor waits for his cue in the wings. His cue is not a clue, for dealing with which, like Sherlock Holmes, he has to get on all fours and study it with a glass. The cue is given, and he leaps straightway into his part. The cue may call up images but more often than not it is a cue for other words, for action, or for seeming passion. In just the same way it is the exception to treat words as clues. Usually this is left to the philologists, etymologists, and lexicographers. We are all logicians, but the symbolic logicians understand their symbols as

we understand words. Having learned the correlations or customary connections by frequent use, they bridge the gaps between the
symbols and other symbols as mechanically as we do between
words and what they mean.

Usually we get many cues at once. This is another way of saying that the understanding of a word often requires that it be
given in a context. Divorced from its context, it is meaningless,
and in others quite different in meaning. A word may suggest
other words which have frequently been associated with it, and,
all together, these may suggest to us the meaning. This illuminates
not only how we understand words but also how we misunderstand
them. Misunderstanding a word is not a case of being mistaken
in what we see or hear but in the meaning we give to it. This has
its parallel in so-called illusions of sense. There are no illusions
of sense, only delusions of the understanding. We are not mistaken
in seeing the crooked oar. We are mistaken if we think it will feel
crooked. Just as Macbeth solved the problem of the dagger before
his eyes by testing whether it cohered with all the other information he could get, so we give the right meaning to a word through
noticing what its context is.

6. Functions of Language

From what has gone before we can collect some of the functions of language. There are probably very many functions; I do
not pretend to know all of them. But for my purposes there are
two which are probably the most important. These I want to
specify.

The first is what might be called the communicative function.
This, according to many, is the most essential function. "The
most essential function of words," according to Lord Russell[1] "is
that, originally through their connection with images, they bring
us into touch with what is remote in time or space. When they
operate without the medium of images, this seems to be a tele-
scoped process." Certainly this is an important function. We use
the words of our language to enable us to think of absent things.
The things we think *of* are not present to thought, only the words
and certain images, and sometimes only the words. But we say

1. *Analysis of Mind* (London, 1921), p. 203.

that we think of these absent things. We do this although some of them, like Peking, the other side of the moon, a chiliagon, and Julius Caesar have never been experienced by us, and, perhaps, never will. Clearly, when words function in this way, they are substitutes for the real things. Often it would be desirable to have the things themselves before us. But there are many advantages in having only the words and images, or only the words. For our thought can range over absent things more rapidly than we can observe them when present; it can capture things already lost; it can make, as it were, things that do not exist; and it can entertain things neither sensible nor imaginable. It is doubtful, however, whether this is the most important function of language. It is essential, but probably it subserves another.

The second function I want to stress is the pragmatic function. Words are used to excite to, or deter from, action. Sometimes words are used to communicate ideas or thoughts. Often ideas are communicated although this is subservient to the author's purpose which is to put the auditor's mind in a certain disposition. Sometimes no thing that can properly be called a thought or an idea is communicated. Just as the actor previously mentioned treats his cue as a direct cue for action "without any ideas coming between," so often do words stir us directly to action. Certainly ordinary language is better designed for operation than for theory. For the former it is admirably fitted; for the latter, not. We encounter difficulty in understanding or in communicating a theory; whereas it is relatively easy to stir or to be stirred to feel or to do something. Certainly when we consider the purposes of the users of ordinary language it seems to have been designed more to satisfy the second function than the first. The main purpose of commercial radio and television, of the speeches of politicians, and of the sermons of priests, is ultimately to induce listeners to do something.

These two functions, both of which are essential functions of ordinary language, may be characterized as follows: words are used as signs *of* things absent in time and space; and words are used as signs *for* action about these things. These functions will have an important role in what follows.

7. Errors in Language

In this section I want to specify some well-known errors connected with language. My reason for presenting material so familiar is that these errors now compose many of the "associated commonplaces" of language. They are, therefore, either features that I shall assume as ingredients of the language model when I use this model to illustrate the nature of vision or features that I need to fill in the background of my illustration.

While the previous account is directly an account of signs and especially of those signs which go to make up language, it is indirectly a catalog of errors of language. I mean by these errors not only mistaken views about language but false beliefs about the world or delusions that we are prone to fall into because of language. Although these two sorts are distinguishable they overlap, because if we have a mistaken view about the nature of words we are likely to have a distorted view of the world, and *vice versa*.

The first way in which we are imposed upon by words is this: We are prone to think that the structure of language mirrors the structure of the world. Most of the mistakes that I shall single out for attention are instances of this. It is now well known. In this century so many philosophers have discussed it that I can be brief. Since Aristotle, ordinary people have had the view that the world is full of things or substances that own properties or qualities. We notice the corresponding fact that most sentences of the Indo-European languages lend themselves to the subject-predicate analysis. This has been the main and obvious way to analyze them. Thus either we have made our language to fit the facts or we have made the facts to fit the obvious structure of our language. What makes us suspect that the latter alternative is true is that there are more languages in existence that are incapable of the subject-predicate analysis than otherwise. The people who speak them do not suppose the world divisible into subjects and predicates. This has been pointed out by Sayce, Russell, Whorf, and others. It shows that one can manage the affairs of daily life without this view, and that the subject-attribute metaphysic is not an innate category. And it suggests that men have chosen this way of viewing the world as they have framed their language: out of convenience.

Secondly, we are prone to suppose that if things have the same name, then they are either the same thing or the same sort of thing, or, at least, have common properties or are alike in some respect. That is, from an identity of name we suppose some identity of nature. But I have tried to show that different things may have the same name although, on reflection, we realize that some are neither the same thing nor the same sort of thing; some have few common properties; some have only the common property (if it can be called a common property) of constantly going together; some, like the things named by metaphors, may have either a remote likeness or none at all until some innovator, like a Pythagoras or a Churchill, tells us it has; and some, like puns, may be alike only in the sound of the name. I say that on reflection we realize all this. But may there not be many cases in which we do not realize it? May we not be partially blinded by the excessive use of names of things which really do resemble each other and which are therefore given one name, so that we cannot see through other shared names, as it were, to different things having no mutual resemblance? It is a well-known psychological law—to which Hamlet appealed—that use can almost change the stamp of nature. The most significant instance of this for my purposes is probably that of mistaking merely constant association for identity.

The third, and the most important, way in which we are prone to error on account of words is connected with metaphor. Metaphors furnish a clear illustration of the "imposture" or the "delusion" or the "cheat" of words. It is part of the nature of metaphor to appear wearing a disguise. This disguise is often so effective that not even automatic brains can penetrate it. When we make a metaphor or when we use one we are either pointing *to* a mere comparison between two very different things or pointing *up* a comparison invented by us. But we do not say this. We speak as if they are the same. The audience has to see through this disguise. For metaphors do not come with their shared names on them. Every case of taking a metaphor literally is a case of being duped by words. And speech is metaphorical more than we imagine until we notice the puzzled looks on the faces of our children who want to feel to see what we are keeping up our sleeves. Every case of a dead metaphor taken in its original sense is a case of being imposed upon by words. We all know that a live metaphor names two different things one of them "not properly," and that a dead metaphor now names this one properly and often the other thing improperly. If I give a dead metaphor its old literal meaning instead of its now literal meaning, for example, if I think that I grasp with my mind what I merely comprehend, I am fooled by words. But if I forget, or am unaware of, the fact that these dead metaphors once were alive—or, if preferred, that these words once were metaphors—I can again be duped.

We tend to forget that there are many subjects that we speak of only in metaphor or, at least, predominantly, for example, the mind and God. The histories of the sciences of psychology and theology record, in large part, the unending search for the best possible metaphors to illustrate their unobservable subjects. We find later theorists objecting to earlier metaphors that, with advances in other sciences, have become obsolete. We find them substituting new metaphors for old ones that are either worn out by over-use or that present an unappealing picture. But we also find many of these theorists writing as if they were replacing false accounts of these subjects by true accounts, or as if they were replacing metaphorical accounts by literal accounts. To take some trivial examples, it may be valuable to replace the metaphor "in the mind," which suggests that the mind is a box or room, by the metaphor "comprehend," which suggests that the mind is active like a grasping hand. But we are fooled by words if, rejecting the former because it is only a metaphor, we then accept the latter because it is not. Similar remarks apply to other replacements, such as that of Plato's political and charioteer metaphors by the *id, ego,* and *super-ego* metaphors.

What I have said about the philosophy of mind applies also to the philosophy of nature. Some of Newton's models are mentioned in his question: "Have not the small particles of bodies certain *powers, virtues,* or *forces,* by which they *act* not only upon the rays of light . . . but also upon one another for *producing* a great part of the phenomena of nature?"[2] It is difficult to discover whether Newton actually thought that there are physical agents that can act or do or produce things; whether he thought that there are real forces residing in bodies enabling them to attract

2. *Opticks,* Query 31, *my italics.*

other bodies; or whether he merely supposed that inanimate things can act or do on the model of persons, just as we now suppose metal fatigue. Whether he did so or not, we deceive ourselves if we reject his account on the ground that, full of metaphors, it is scientific mythology, and accept another on the ground that, literally true, it is *bona fide* science.

Fourthly, we are cheated by words if we take every word as a name. But there are many related errors. Until Lord Russell exposed an absurd mistake,[3] it seems that many thought that every definite description names a thing. Meinong, for example, at the turn of the century, held that such phrases as "The Golden Mountain" named entities that exist in some strange way because he could ask, after admitting the non-existence of the Golden Mountain: "What is it that does not exist?" And Russell himself admits that, before he hit upon his new analysis of descriptive phrases, he too was inclined to give a sort of shadowy existence to the supposed referents of such phrases. A similar mistake is made if we suppose that every common or abstract noun has a precise designation. This, it seems, is the cause of the belief in the existence inside or outside the mind of abstract ideas or universals. The question of the nature of universals remains a controversial subject. Since no one can find universals no matter what various guises they wear: "the meaning," "the connotation," or "the common property," either in sense, intellect, or imagination, it seems that their reification is due to the mistake I have named. Students learn from text-books to distinguish between the denotation and the connotation of a word. The former lists things. But the latter purports to name the common characteristic(s) of a class of things, without which, things would not be named as they are. Moreover, the connotation is named by the right-hand side of a lexical definition. It is the meaning of the word. How easy it is to suppose either that this defining property or meaning or connotation exists apart from particular things or that we can entertain it as a precise, discrete idea in our minds. There is a great temptation to be fooled because so many of the words of our ordinary languages look more like proper names of discrete entities than like shared names of many individuals. It is easy to suppose that there

3. "On Denoting," *Mind* (1905).

are intermediaries between common nouns or abstract nouns and particular things. It is certainly convenient to speak *as if* there are—*as if* there are such things as Force, Disease, Heredity, and Democracy. But the line between make-believe and belief is thin. The mistake made is analogous to that made when we are victimized by metaphors. Nobody wants to relinquish their use, even if this is possible. The ideal is to continue to use them with awareness. We have seen that those philosophers who wrote about the mind were searching for the best metaphors to illustrate it. But few were aware of what they were doing. Most were the dupes of metaphors previously made or of those of their own devising. For example, it seems as though Locke mistook his model of the mind, at least in part, for the thing modeled: an empty room at birth which was slowly filled with ideas.

Fifthly, and corollary to the last, we tend to suppose that sorts or kinds or classes are determined and set out by nature. And what goes for sorts also goes for numbers. I have tried to argue that sorts and numbers are nothing fixed and settled, being imposed by us largely out of convenience. To suppose otherwise is especially easy because, on the face of it, our common names like "dog," "man," "tiger," etc. seem to name sorts intervening between them and individuals; and because most of the sorting has been done for us by our ancestors and fixed for us by dictionary definitions. Sorts, it seems, are either fixed by nature or, as it were, built into the structure of our minds. That neither is the case is evident when we reflect that most of the arguments about sorting are verbal, that it is merely a matter of convenience and convention whether we put certain individuals into this or that sort.

Sixthly, we are prone to treat constant associations as necessary. This is in part a case of being imposed upon by words and in part not. We are exceedingly prone to imagine that things have in their own nature a fitness to go together or to succeed each other as they do independently of our noticing that they happen to do these things. We think that they must be connected as they are. It is certainly difficult to imagine the sounds, smells, looks, and feels, that we now experience as constantly co-existing or succeeding each other in various ways, as occurring otherwise. It is not too easy to imagine the look of an apple being followed by the texture and taste of an orange; a dog's head appearing with a giraffe's body; an ordinary plane mirror image being inverted as well as "sideverted"; our common words without thinking of their meanings; or the looks of things without their customary feels sensed or imagined. Although we realize that these associations might have been otherwise, we are prone to forget that they have been built into our minds by slow experience. Our proneness to forget it is due in part to the almost exceptionless co-existence or repetition of events, and in part to the power of words over us. Every connection need not be what it is and could be another thing. But we tend to export the necessary connection that we have made between words to the connection that we only find between matters of fact. This error has been so often exposed since Hume that I need not dwell on it.

Lastly, there are two mistakes regarding the understanding of words that have been so often exposed that I shall do little more than mention them. One is that although we may learn to understand many words by building up associations, we must not think that every time we understand a sentence we must form a mental proposition answering to the verbal, whether this proposition is taken to be either a content of intellect or imagination, a conglomerate abstract idea, or a series of images. We understand the sentence without "cashing" its words in things or in images although we can often do this and often do. Since there are many words that cannot be treated thus, for they denote nothing yet are understood, it is clear that we learn to understand them by using them in contexts. The other is that we must not accept the notion that, in order to understand the words of a language, we must know the grammar of them and their dictionary definitions. That is, we must distinguish between *knowing how* to speak or listen to words and *knowing that* about them. Learning a language is primarily the former.

We need hardly hope that we can be entirely delivered from the deception of words merely by becoming aware of the errors I have described. We need constantly to be on guard against the snare of shared names which might easily lead us to believe that things very different are the same. We need to guard against being deceived by the disguise metaphors wear—a disguise which often increases in subtlety with increasing age. We need to penetrate also the disguises worn by common nouns and abstract nouns, realizing that, although they appear to name sorts and universals, there are no sorts in nature and no generals either. We need to guard against the tendency to export the necessary connections, made by us, from words to things. And we need to guard against assuming that language mirrors the structure of the world. To help us in this last and other similar cases I welcome Lord Russell's cure: "Against such errors," he says, "the only safeguard is to be able, once in a way, to discard words for a moment and contemplate facts more directly through images. Most serious advances in philosophic thought result from some such comparatively direct contemplation of facts."[4]

It may seem to the reader that the errors I have marked out for attention are too obvious to be worthy of mention; that, at any rate, he has long since managed to avoid them. If so, he will agree, I hope, that these, if they are not errors, are at least pitfalls for the unwary and the unaware.

4. *Analysis of Mind*, p. 212.

CHAPTER V

Visual Language

1. The Problem of Vision

IN DESCRIBING the nature of language I have used other things to illustrate it. This is a familiar practice. Wittgenstein once said that sentences get their meanings by being pictures of the world. Later, more appositely, he likened words to tools in a tool box. Hobbes rejected the "money" metaphor in favor of "counters": "Words are wise men's counters; they do but reckon by them; but they are the money of fools."[1] To treat words as counters at a card table cashable at the end of the game corrects the mistaken view that, in order to understand words, we must "cash" them as soon as they are uttered. Some of these illustrations paint distorted pictures. Some, like that of tools and counters, are vivid and helpful. Many may think, however, that we know well enough what a language is and that we do not need philosophers' models to help us understand it. Its main features are so plain that instead of using other things to illustrate it, language itself may be used to illustrate other things. For example, in a famous passage in *Il Saggiatore* Galileo used language to illustrate the physical world:

> Philosophy is written in that vast book which stands forever open before our eyes, I mean the universe; but it cannot be read until we have learned the language and become familiar with the characters in which it is written. It is written in mathematical language, and the letters are triangles, circles,

---

1. *Leviathan,* 1.4. Professor H. H. Price, *Thinking and Experience,* p. 239 ascribes an improved version of this epigram to Bacon: "Words are the counters of wise men, but the money of fools."

and other geometrical figures, without which means it is humanly impossible to comprehend a single word.[2]

This passage has many interesting features. The language of nature is a visual language since it can be read by using our eyes. But it is also a mathematical language, for geometrical figures are found in it. Nature can be deciphered, not by the plain man but only by the expert in mathematics. The passage thus reveals, at the dawn of the modern epoch, the attitude of Galileo and his modern followers to nature.

But there is something wrong with this interpretation of the metaphor when applied to vision. For it is not true that nature is a closed book to most of us although, in Galileo's account, it might just as well be. His "book" may indeed lie forever open on display at the circulation desk but it might just as well lie closed and locked in the treasure room. There should be other interpretations of the metaphor that will let us use less esoteric ways of reading the language. For we all know how to read the book of nature although we may not understand the grammar of it. We need no university training in deciphering Galileo's language to enable us to read the books in the running brooks or to listen to the sermons in stones. Children and illiterates are able to understand the language although they cannot read it. Accordingly, an interpretation is needed that will allow the book to be, as it were, read aloud to us.

Let me then reinterpret it. Now that we are aware of the main features of the language model, let me apply it to one particular subject, vision, the most comprehensive of all our senses. If it will enable me to account for many of the facts, its utility will be confirmed. If it will enable me to account for some facts unaccounted for on other theories, can do this more economically, and can suggest new techniques, its relative value will be increased.

A good metaphor sometimes enables us to learn not only more about the nature of the thing illustrated but, through it, more about the nature of its literal meaning. Accordingly, through a detailed study of vision we may learn more about language. A

2. *Il Saggiatore*, Q.6 (*Opere*, vi), 232, quoted from A. C. Crombie, *Robert Grosseteste and the Origins of Experimental Science* (Oxford, 1953), p. 285.

good metaphor is also a beguiling thing. Once it is understood and accepted, one sees the thing illustrated through new spectacles that, when worn for a while, are hard to discard. The iron curtain between the East and West will either wear out or remain until a new metaphor is made; but the attraction once merely supposed to exist between all bodies in the universe now seems actually there. This metaphor has entered the very marrow of our bones and seems impossible to remove. Once understood and accepted, some may find it equally hard to discard the language metaphor.

The problem of vision to which I address myself was set by the Greeks. It begins with the consideration of certain popular suppositions, the main one being that when we see we are directly or immediately aware of physical objects at various distances from us, of various sizes, shapes, and positions, physical objects that we can sometimes touch, hear, smell, and taste, as well as see.

This supposition admirably accommodates ordinary cases of vision. But it meets difficulties in many that are not so extraordinary. Illusions, hallucinations, cases connected with the relativity of our perceptions, and certain scientific discoveries, present problems. The senses, it seems, often deceive us. Macbeth saw what he took to be a dagger but could not clutch it. What, then, did he actually see? We see what seems to be a lake before us in the desert, but we dip our pannikins into sand. We see what could be a yellow cup, but we realize we look with jaundiced eyes. My thumb looks bigger than the Eiffel Tower, but I do not think it is. We see from here a small round tower in the distance, but when we get there we climb a large square building with battlements and turrets. We see a bent stick partly submerged in water, but we pull out a straight one. If we look through a magnifying glass at an object out of focus as we move the eye backward we notice the object getting nearer and larger, but we know that it is getting farther away. The moon looks bigger at the horizon than at its zenith, but we do not think it really is.

Cases like these, whether we are genuinely deceived or not, prompted Aristotle and most subsequent scientists interested in optics to ask such questions as: "How can what we sense be thought identical with the physical object when the latter remains

fixed though the former varies?" Their answers persuaded many
of them to give up the popular supposition of Direct Realism
(though many others have tried valiantly to retain it), and to
distinguish what we sense from what we perceive.

It will be helpful to consider my account as an attempt to solve
the problem set by Aristotle in his *De Anima*. Aristotle's own account, however, is ambiguous. My interpretation is as follows.

What we sense are sense-data, the proper objects of each sense.
"By a proper object I mean one that cannot be sensed by any
other sense and in respect of which no error [or truth] is possible.
Thus color is the proper object of sight, sound of hearing, and
flavor of taste; while touch, on the other hand, has several proper
objects, . . . viz., hot cold, dry moist, hard soft."[3] In receiving
these sense-data the mind is purely passive, being affected "just as
a piece of wax takes on the impress of a signet ring."[4] The mind
is also undeceived, or, rather, neither deceived nor undeceived, for
the sense-data are "in the mind just as characters may be said to
be on a writing-tablet on which as yet nothing actually stands
written."[5]

What, then, do we perceive? In addition to the proper objects
there are objects "common to all the senses," such as size, number,
figure, movement and rest. "There are movements, for example,
perceptible to touch as well as sight."[6] But these common objects
are not sense-data because there is no special sense organ for them.
"All of them are sensed *indirectly* as a result of the functioning
of the particular senses."[7] There are, moreover, other objects perceived, which Aristotle merely described as "indirectly sensible
objects," e.g., "we see that the white object is the son of Diares."[8]
If the mind is passive in receiving the sense-data, then it is active
in perceiving the common and indirectly sensible objects. According to my interpretation, therefore, if what we sense are *sense-*

3. *De Anima,* 418a, 423b.
4. *De Anima,* 424a.
5. *De Anima,* 430a.
6. *De Anima,* 418a.
7. *De Anima,* 425a. Elsewhere Aristotle speaks as if the common objects are
directly sensed; e.g. at 418a, but immediately he admits that only the proper objects "are sensed strictly speaking."
8. *De Anima,* 418a.

data, then what we perceive are appropriately regarded as *mind-facta.* If the sense-data record no lies (and no truths either), then in perceiving we may err. For example, "sight is infallible in its awareness that a certain visual datum is white, although it is perhaps deceived in taking this white datum for a man. . . . For falsehood [and truth] always involve a synthesis."[9] If the sense-data are merely characters on Aristotle's writing-tablet, then I can infer that the things perceived, the mind-facta, are these characters synthesized or composed into sentences. For only sentences can be true or false.

In this fashion Aristotle solved the problem of vision. Seeing is not a simple and direct sensing of physical objects; it is rather a complex conceptual act. It is like making an assertion that something is the case. Some of our assertions are true; others false. Thus Aristotle solved the problem of visual illusions. There are none. There are only delusions or false beliefs of the active mind. But this solution created a bigger problem. What principle enables us to bridge the gap between the sense-data and the physical object? Aristotle's answer, "the central sense or general sensibility,"[10] left nearly all to be done. Not quite all because he gave a hint at a most satisfactory solution. The hint he gave was that the sense-data are the elements of a language.

2. The Linguistic Solution

Once I make believe that vision is a language, I can apply as many features as I need of the latter to the former in order to illustrate how we see. In what follows I shall exhibit some of these features and show how they may be interpreted to illuminate the facts. When appropriately interpreted they will form the rules or principles of a theory of vision. The linguistic model I have in mind is an ordinary language like English or Italian, already in existence, and, more often than not, it is the spoken rather than the written language.

Language is the conventional use of signs functioning not only and not chiefly to communicate information but also to arouse

9. *De Anima,* 430b.
10. *De Anima,* 425a.

emotion and direct action. Accordingly, the signs of an ordinary language like English and, therefore, of visual language function as *signs of* things or as *signs for* action or passion. When uttered or presented they may make us think of other things absent or distant from our bodies in time and space, or they may make us leap straight into action or passion. If telling what words mean is interpretation of signs, so is seeing things in space; and to see must largely be to foresee so that we may take action.

But what sort of things do we see? The obvious answer, already given in the last section as part of the main popular supposition, is that we see physical objects in space, some nearer, others farther off, of various sizes, shapes and situations, physical objects that are colored, hot or cold, rough or smooth and that sometimes have taste, smell, and resonance. But by using the linguistic model I decide at once that these physical objects are not directly sensed by sight. For when we listen to someone talking about rocks and trees we do not hear the rocks and the trees. We hear various sounds that may make us think of rocks and trees. These sounds function as signs of the things they signify to us. Using this feature of the model, I draw a similar distinction in seeing. Accordingly, physical objects are seen indirectly or perceived by sight. They are at least some of the things signified by the signs of visual language.

In order to present my preliminary solution to the problem of vision as clearly as possible, it is necessary to simplify this account of the things signified. This I do by abstracting some of the features, just mentioned, of physical objects and ignoring, for the time being, the rest. Thus I assume that physical objects are those objects commonly regarded as the spatial properties of objects, specifically, depth or distance, size, shape, situation, and movement or rest. These objects, I assume also, could be known by us if we had not been gifted with sight. This is not an implausible assumption for it appears that congenitally blind people manage to acquire spatial awareness.[11] Since such awareness must be dependent

11. Nevertheless, it has been questioned. Cf. M. von Senden, *Space and Sight* (Glencoe, Illinois, The Free Press, 1960, tr. by Peter Heath from the German edition, 1932), p. 309. Von Senden writes: "We have been led to conclude that by tactual perception alone the patient is unable to acquire an awareness of space, and that this is solely dependent on visual perception."

ultimately upon the tactual sense, I shall often call these objects "tactual objects." It will be noticed that these objects are much the same as Aristotle's common objects.

What, then, are the signs? With the help of the model I decide at once that just as sounds, variously combined, become signs in a spoken language so do colors of various hues and intensities, variously combined and ordered, become signs in this language of vision. In similar fashion, also, they may be supplemented by data of the other senses. Thus when we see we may use as cues such feelings as eyestrain as well as those that accompany the minute turn of the eyes and the more massive movements of the head.

Equipped with these interpretations, I can now address myself to what is perhaps the most vexing problem of vision: What is the connection between visual data and physical objects? I have already suggested that one popular solution, now fashionable among philosophers, meets difficulties in accommodating illusions, hallucinations, etc. This solution, that of Direct Realism, is that the connection is one of identity: we sense and perceive the same thing. We shall see later that another solution, widely held among scientists and philosophers from the time of Kepler and Descartes, meets similar difficulties. This solution, that of the Representative or Copy Theory, is that the connection is such that it enables us to make inferences to physical objects, inferences that are based either upon likeness or upon geometrical necessity. What principles, then, can I appeal to that will enable me to give an adequate solution to the problem, one that will accommodate the ordinary facts as well as the not so extraordinary facts connected with illusion? These principles are conveniently found in the features specified in the linguistic model when properly interpreted.

The connection that I need is the same as that found between the signs and the things signified in a language, between words and their referents. But first, what is it not? In the following three rules I deprive myself of any recourse to an identity or picture or deductive theory of language, and thus to a direct realist or representative (including geometrical) theory of vision.

I The sounds or colors that become signs of a language are neither identical with, nor pictures of, nor necessarily connected with, the things they signify (Cf. *E*, 127; *V*, 41).

For example, the sounds conventionally named by the marks C A T S or the marks G A T T I are neither identical with nor images of cats, nor can we deduce cats from them. The two sorts of things are, indeed, as specifically distinct or heterogeneous as the smell and the taste of an onion or as the look and the taste of cheese or whiskey. Moreover, these sounds and marks might have been sounded or written otherwise. Men might have used the marks D O G S or G I R A F F E S to denote cats. This is true because:

II If the sounds or colors that become the signs of a language were identical with, or pictures of, or necessarily connected with the things they signify, then we could interpret them although none of them had been ostensively defined for us (Ibid.);

and

III The consequent of Rule II is false; that is, we could not interpret them unless at least some of them had been ostensively defined for us (Ibid. Cf. *A*, IV. 11).

An ostensive definition consists in the establishment of an association through the hearing or seeing, etc. of closely similar sounds or colors, etc. whenever the object to be defined is present.[12] The establishment of such an association requires time and experience and repeated acts, so that we acquire a habit of knowing the connection.

How can I test this part of the theory? I deduce that a foreigner to a language cannot understand it. This is easily tested in the case of ordinary languages; most of us have tested it on numerous occasions. A foreigner to Italian, for example, upon first hearing the sounds **gatto** and **kařne** would not think of the things signified by them although he may know all about cats and dogs. Similarly, since I have decided that vision is a language, that is, that seeing things is nothing but understanding a language, I deduce that a foreigner to visual language cannot understand it. For example, a congenitally blind person now adult who is suddenly made to see, upon first "seeing" certain colors, would not think of cats and

12. This definition slightly modifies Bertrand Russell's in his *Human Knowledge* (New York: Simon and Schuster, 1948), p. 501.

dogs or any other objects in space although he may know what they are by touch. He would prove quite unable to recognize what they are or to name them. He would get no *meaning* from the spinning mass of colors before his eyes. That is, he would not be able to see.

This brings me to one of the most celebrated *Gedankenexperimente* in the history of philosophy, namely, the Molyneux problem,[13] which Locke published as follows:

> Suppose a man born blind, and now adult, and taught by his touch to distinguish between a cube and a sphere of the same metal, and nighly of the same bigness, so as to tell, when he felt one and t'other, which is the cube and which the sphere. Suppose then the cube and sphere placed on a table, and the blind man to be made to see: *quaere,* whether by his sight, before he touched them, he could now distinguish and tell which is the globe, which the cube?[14]

The Molyneux problem became a central problem in eighteenth-century epistemology and psychology.[15] It is still of importance today, for the problems it raises have not yet been solved.[16]

13. William Molyneux (1655–98) published his *New Dioptrics* in 1692. He became a close friend of Locke in the nineties and introduced Locke's *Essay* to Trinity College, Dublin, before Locke was known at either Oxford or Cambridge. He sent his "jocose problem" to Locke on March 2, 1693: "I have proposed [it] to divers very ingenious men, and could hardly ever meet with one that at first dash would give me the answer to it which I think true till by hearing my reasons they were convinced." On March 28, 1693 Locke replied: "Your ingenious problem will deserve to be published to the world." On December 24, 1694 Molyneux sent to Locke a copy of a letter from Mr. Synge to Dr. Quayle showing "by what false steps this gentleman is led into his error," the main false step being the "same name" of the two objects. See Locke, *Some Familiar Letters* . . . (1706), from Locke's *Works, 3,* 512, 514, 542, 546.

14. *Essay,* II. 9. 8.

15. Cf. Ernst Cassirer's remark: "A survey of the special problems of eighteenth-century epistemology and psychology shows that in all their variety and inner diversity they are grouped around a common centre. The investigation of individual problems in all their abundance and apparent dispersion comes back again and again to a general theoretical problem [the problem of Molyneux] in which all the threads of the study unite" (*The Philosophy of the Enlightenment,* Princeton University Press, 1951) first published in German, 1932, p. 108.

16. Cf. M. von Senden, *Space and Sight.* Von Senden says, p. 309, that his book "took its starting point from the celebrated problem of Molyneux."

Both Locke and Molyneux answered "No" for the reason that the blind man has no experience that an angle in the cube that pressed his hand unequally appears the same to his eye. Leibniz[17] and a host of others[18] answered "Yes." Leibniz maintained that because shape is common to sight and touch, the once-blind man could recognize the cube by "the principles of reason." He argued that we all possess "a natural geometry," and that the feat could be performed by dint of "reasoning about rays according to the laws of optics." Is there a common nature? "Is there an inner connection which permits us to make a direct transition from one such field to another, from the world of touch, for instance, to that of vision?"[19] If there is such an inner connection; if, for example, we see the same size and shape that we touch,[20] or the two objects are no more unlike than ⊲▢ and the dog it pictures,[21] or they are necessarily connected in the same way as the real image and the object in a geometrical construction,[22] then a blind man on first acquiring sight would be able to recognize the cube and the sphere. Certainly Leibniz and many others thought that he would be able to do so for reasons such as these. Indeed many people whom I have asked agree. This is Rule II interpreted. But if the once blind man cannot recognize them (Rule III), then Rule I is true. In other words, either Rule I is true or my solution to Molyneux's problem is wrong. This solution is easily reached once I make use of the rich offerings of the linguistic model. The blind man at first sight of the visual cube is in the same plight as that

17. *New Essays*, 11. 9. 8.

18. For example, the mathematician Jurin in Robert Smith's *A Compleat System of Opticks* (1738), paras. 160–61.

19. Ernst Cassirer, *The Philosophy of the Enlightenment*, p. 108.

20. This characterizes the Theory of Direct Realism. Cf. Descartes: "That bodies move, have various sizes, shapes, and motions . . . ; these facts are observable not just by one sense but by several . . . ; the same does not apply to other sensible qualities such as colour and sound; they are not observed by several senses, but each by one sense only" (*Principles, IV.* 200). Descartes was not a Direct Realist, but this assertion can be considered apart from his main doctrine.

21. This characterizes the Representative or Copy Theory. Cf. Locke: "The mind knows not things immediately but only by the intervention of the ideas of things it has of them," and visual data are "visible resemblances or ideas of things without" (*Essay*, II.11.17 and IV.4.3).

22. This characterizes the Geometrical Theory. Cf. Leibniz, text to note 17 above and Ch. VII below.

other foreigner who, for example, hears the sounds **gatto** for the first time. Nothing that either foreigner sees or hears can suggest to him the cube or the cats he knows so well.

But surely my solution can be confirmed or disconfirmed by actual tests on people blind from birth and made to see. Several tests have in fact been performed since Molyneux posed the problem. Fortunately many of the surgeons' reports have been collected and analyzed by M. von Senden.[23] In the operation for congenital cataract, the opaque crystalline lens is removed and replaced by an artificial lens usually in front of the eye. Before the test commences the surgeon must be confident that visual acuity has been achieved and that the patient has overcome his postoperative intolerance to light. Of the cases reported none, I think, disconfirms the answer given by Locke and Molyneux. I shall not here enter into the details of the voluminous reports on these tests. It will be enough if I give portions of three typical reports on tests upon patients who had been prepared before the test by tactual trials with the "same" objects, as Molyneux had prescribed:

> He is shown a sphere and a cube, both of the same colored wood and of similar cross-section. On looking at them together he realizes that the two are distinct, but *does not know which is round and which cornered.*[24]

> He could at once perceive a difference in their shapes; though he *could not in the least say which* was the cube and which the sphere, he saw they were not of the same figure.[25]

> "Do you know what a square is?"—He positions his two hands so that they form a pair of surfaces which make contact almost at right angles along the radial edge. He thereby produces an angle, which is actually part of a cube. "And a circle?"—He again bends his hand round with the fingers pointing towards the wrist, and thereby produces an almost complete ring. After this fashion he therefore has some knowledge of circu-

23. *Space and Sight.* He lists eighty-three cases and discusses sixty-seven.
24. Raehlmann's case, 1891, quoted in *Space and Sight,* p. 114, my italics.
25. Nunneley's case, before 1855, quoted in *Space and Sight,* p. 106, my italics.

larity. In looking at the watch, at which his gaze is obviously directed, he remains absolutely *incapable of saying whether* it is round or cornered. However much I insist on an answer, none is forthcoming. . . . On the following morning the same question; the same inability to answer. So I then let him feel the watch. No sooner has he taken it in his hand than he immediately says, "That's round, it's a watch." [26]

Von Senden himself concludes that "amid the initial chaos of blurred and vibrating colors, the patient is at first unable to make out anything definite at all," that "the patient's visual field contains nothing beyond a set of perfectly genuine visual objects, which are still devoid of any significance," and that "these visual impressions awaken no familiar ideas in him, and that he cannot recognize the objects in question." [27] This conclusion, it seems to me, unlike some others of von Senden's[28] is a correct inference from the evidence.

Can the patient recognize objects placed in front of him? Yes, if he can name them and perform appropriate actions observable by spectators; no, if he cannot do these things. We need no combination of Locke, Leibniz, and Newton to frame the questions for this part of the test, namely, that of recognition by means of visual objects or of understanding of the words of visual language. The questions have been adequately framed and the solution given in favor of that offered by the great original who posed the problem.

26. Dufour's case, 1875, quoted in *Space and Sight,* p. 108, my italics.
27. *Space and Sight,* pp. 295, 299.
28. For example, his views that awareness of space is purely visual and that the blind have no awareness of space: "It is by operation alone that the congenitally blind can achieve an awareness of space" (p. 289). The latter view is not, I think, supported by the facts presented in his book. But he argues for it as follows (p. 66): If the congenitally blind were aware of space, then, on being made to see, they would recognize objects in space; and this they cannot do. The first premise would be true, I think, only if we add to it the assumption that the fields of sight and touch have as much similarity as I supposed in my Rule II. In which case we get an entirely different conclusion. It is as if von Senden concludes that a Chinese unfamiliar with English does not know what trees are, because if he did, then he would understand the sounds tʀiyz the first time he heard them. But he could do this only if tʀiyz and trees were alike.

Having divorced myself from all identity and picture theories of language and, accordingly, with the help of the linguistic model, from all Direct Realist and Copy theories of vision, I have not yet solved the main problem: How are the signs of visual language connected with the things they signify? In the next three rules I offer the main ingredients of the solution to the problem set by Aristotle, and in doing so I exhibit still more of the extraordinary illuminating power of the linguistic model. I do not pretend that the principles I give here are enough to explain how we understand a language. But I think that I can collect the main ones.

IV   If some similar sounds or colors have been ostensively defined for us, then we can interpret them, and we find it increasingly easy to interpret other sounds or colors that are new to us (Cf. *PC,* 225; *E,* 25; *V,* 47) .

This slightly modifies the converse of Rule III. It is the main way in which we understand words. Thus the basic terms of a language are learned by the constant and long association of entirely different things. After the auditory or visual data are ostensively defined by repeated acts, he who was once a foreigner to the language will consider them as signs and will be able to tell what they signify. The process is arduous and irksome, the achievement difficult in languages either ordinary or visual. It takes months, or even years, for a foreigner to learn to understand sounds or colors.

When the basic terms of an ordinary language have been learned ostensively (although most of our common words *are* learned in this way), we can interpret others by defining the latter in terms of the former. In the case of visual language we rely on analogies stored in our memories. Consider the last report I gave on a once-blind patient.[29] Here is a person tortuously learning by ostensive definition to interpret the visual words that signify a round object and a watch. The act described will be repeated on numerous occasions. In time he will learn to tell by sight what a round object is. Eventually, with the help of analogies, he will even be able to see the moon and, possibly, flying saucers, although he has never

29. See text to note 26, above.

handled these objects. Similar remarks apply to the acquisition of the whole of his basic visual vocabulary containing the visual analogues for "square," "angular," "straight," "curved," etc.[30] Many of our visual words are "cashable," in things, although it becomes increasingly rare for us to "cash" them every time they are presented. At this stage, then, the "money" metaphor gives way to that of counters at a gaming table cashable at the end of the game (Cf. *A*, VII).

All this requires that the foreigner have a memory enabling him to make inductive leaps. The analogies he draws on influence his interpretations. Accordingly,

V Our expectations or prenotions regarding the size, shape, and situation of things condition the meanings we give to the words we see or hear (Cf. *V*, 59).

Interpretation, whether in reading a page, listening to a lecture, or seeing a landscape, is always an active process conditioned by what we expect. To a marked extent we see what we are looking for because we are thinking of it. Man Friday looked and saw nothing, but Crusoe looked and *saw that* it was a ship.[31] Macbeth saw a dagger, and Hamlet the ghost of his father.

Nevertheless we often misinterpret signs. But both the sources of our error and the means for its avoidance are conveniently specified in the linguistic model. First, the sources:

VI(a) Words are often ambiguous, that is, they do not "always suggest things in the same uniform way and have the same constant regular connection with matters of fact" (*A*, IV. 7).

Without the means whereby we might avoid being taken in by such ambiguities, commonly called in English language visual illusions, "we should no more have taken blushing for a sign of shame than of gladness," or a bent appearance for a sign of a straight stick than of a crooked one (Cf. *E*, 65).

---

30. Cf. von Senden, *Space and Sight*, p. 304: "It becomes increasingly simple to perceive shape, not only in familiar objects, but also in those that are new. The patient finds fewer and fewer objects for whose structure he cannot find analogies of some kind in his memory, which guide his interpretation in a definite direction."

31. I owe this example to Roderick Chisholm.

VI(b) Many words are meaningful although they denote nothing
       (Cf. *E*, 64).

These signs can never be "cashed" in things on demand because
they are not even "cashable." But we do not thereby deprive them
of meaning. For example, "we may sometimes perceive colors
where there is nothing to be felt" (*E*, 103). Without the means
whereby we might avoid being fooled by these non-denoting
expressions which, in visual language, are exampled by rainbows,
mirages, certain mirror and lens' images, and hallucinations, we
should cease to find any value in vision. What, then, is the means
whereby we may avoid falling into error?

VI (c) The means whereby is the context of these expressions, for
        a word used in one context or circumstance often has a
        different meaning when used in another context or cir-
        cumstance (*E*, 73).

This rule, properly interpreted, enables us to solve the problems
created by the frequently encountered cases of ambiguity and
absence of denotation of the words of visual language. It is of
extreme importance, for it shows how the linguistic model may
suggest an appealing solution to that problem from which the
problem of vision itself took its start, namely, the problem of illusion. At this point, then, both the "money" and "counter" metaphors, used to illustrate language, give way to the "tool" metaphor.
Only after long experience of the different uses of words within
different contexts and the remembrance of them are we able to
avoid being victimized by the metaphors, other ambiguities, and
incomplete symbols of visual language so that we can *see that*
such things as sticks, crooked in the *context* of water and air,
are really straight, that the blush that suffuses a cheek *means* gladness and not shame, and that the dagger-looking object before our
eyes is to be correctly interpreted as "a dagger of the mind" and
not a real one. *We* are able, and so it is amusing to watch monkeys
being fooled by the ambiguous language of the plane mirror. In
these examples we cannot say that the appearances of the stick, the
blush, and the dagger are either erroneous or true. We might just
as well say that the words "stick," "blush," and "dagger" are true

or false. But we can say that the sentence "This is a dagger of the mind" is.

I have not yet given all the main ingredients of the solution to the problem of vision. With the first three rules I showed how the gap between the words of vision and what they mean is *not* bridged. With the second three I showed how the gap is *partially* bridged. This is still not enough. For most people do not think that there is any gap to be bridged at all. They ordinarily say and believe that the cup they see is the same thing as the cup they feel. They say and believe that they see *one* thing, the physical object, which has the qualities of size and shape, as well as color, smell, taste, and sound, and that some of *its* qualities can be both seen and felt. Moreover, it may be objected at this stage that ordinary people do not think that they venerate the tactual object as much as it might appear that I have been doing, and that I am forced to give the tactual object such an important role in order to provide visual words with their requisite nominees, thus debarring me from applying to learning to tell by touch what I apply to learning to tell by sight.

Accordingly, I must show how the gap is *completely* bridged. I must offer an adequate explanation of how we manage to see the one physical object. If I cannot do this, then, although I may have shown how to correlate sight and touch, I have not solved the whole problem. But if I can show how the objects of sight and touch are interwoven or concreted together to make one thing, then it should be easy to do the same thing with the objects of all the senses. In this fashion that set of beliefs known as the Theory of Direct Realism should be accounted for by the Linguistic Theory. The linguistic model is rich in its offerings here. It offers a just and ready solution. My solution, presented in the next three rules, will be in its terms.

VII (a) We sometimes want to talk *about* a language as well as *from within* it. When this is done it is customary to call the signs and the things they signify by some of the same names (Cf. *E,* 140; *V,* 45; *H,* III; etc.).

This custom is highly convenient. If we were not to adopt it, the endless number or confusion of new names would render language impracticable. Thus when we use a spoken language, it is customary to *call* written words and the things they signify by the same name. We use, for example, the phonetic sequence ɪŋglɪʃ to name both the written and the spoken languages, the phonemes ɛʏ to name the character "a" and its corresponding sounds, and the sounds skwɛ:ʀ to name both the figure square and the six characters used to mark it. Similarly, in the written language of music, a mark on a page is used to name a sound. But in the spoken language of English both the mark and the sound are called by such same names as **nowt** and **haʏ** or **low**. Similarly, in the traditional language of painting, whose words, like those of visual language, are color combinations, a certain combination is used to signify a situation. But in English language both are called by such names as those written "landscape," "view," or "still-life."

Similarly, in the language of vision a certain combination of colors signifies a certain large, high, square tower a long way off. But in English both this sign and the thing it signifies are called by the same names as those written "large," "high," "square," and "far"; the same names as those used by a blind Englishman who has walked to and explored the tower. But why does the thing signified so often confer its name on the sign rather than the converse?

VII (b) Since signs are little considered in themselves or for their own sake, we often overlook them and carry our attention immediately on to the things signified,[32] where nearly all our interest lies (*A,* VII.12).

Our interest lies more in tactual objects because sticks and stones may break our bones but names will never hurt us. Hence Pavlov's dog was more interested in his food than in the sound of the bell,

---

32. Cf. Locke, *Essay,* II. 9. 9: "A man who reads or hears with attention and understanding . . . takes little notice of the characters or sounds but of the ideas that are excited in him by them"; and also Descartes, *The World,* ch. I: We understand what words signify "even without our paying attention to the sound of the words or to their syllables."

while we are more interested in the feels of precipices and port-
manteaux than in their looks.

VIII But these common names do not name common natures or
abstract ideas.

What, then, is the nature of such shared naming? It is what I have
called "sort-crossing," [33] defined as the representation of the facts
of one sort in the idioms appropriate to another, or, in Aristotle's
words, as "giving the thing a name that belongs to something
else." Such sort-crossing may be grounded upon resemblance, but
it need not be. Every case of sort-crossing is potentially a metaphor. Indeed it always is a metaphor in the etymological sense of
"metaphor." Thus we speak of a "higher" and a "lower" in the
notes of music, and we say that men speak in a "high" or a "low"
key. These are no longer commonly regarded as metaphors, but
they can easily become so for him who entertains the two meanings at once. On the other hand, when we speak of "cold" shoulders or "bitter" words, there is no need to resuscitate the metaphors
by thinking of cold shoulders of mutton in the refrigerator or bitter almonds in the mouth. But the line between "high" and "low"
notes on the one hand and "cold" shoulders and "bitter" words
on the other is thin. Thus when an original "sort-crosser" decides
that certain marks above the line on a musical score and certain
sounds are to have the name "high" in common, he need find no
common nature or abstract idea *height,* or, indeed, any resemblance between the height of a mark on a page and the "height"
of a sound; although his decision may create the illusion of resemblance (Cf. *V, 46*).

So likewise we often want to talk *about* the words of visual language. When we do so, such words as "high" and "low," "up" and
"down," "left" and "right," etc., are made use of. This, too, must
be a case of sort-crossing from the things signified to the signs. Obviously this applies not only to the situation words of an ordinary
language but to distance words, size words, and motion words, if
they are used to talk about the words of visual language. Once
more it can be said that this custom is highly convenient. If we had
not adopted it the endless number of new names would hopelessly

33. See above, Ch. I, sec.

clutter up our language. We have even extended this custom so far as to say that smells are high and that ice looks cold, water wet, and surfaces rough. Once more we can say that we need find no common nature between the fixed size and shape, etc., of the physical object and the fluctuating sizes and shapes of the visual signs, nor indeed any resemblance between them, although we give them the same names[34] (Cf. *V*, 46).

IX But if the ostensive definitions of a language are, as it were, sucked in with our milk so that we cannot remember having learned them; if this language is universal; and if in our native tongue we call its signs and the things signified by the same names; then we cannot avoid confounding the signs with the things signified, and we suppose an identity of nature (Cf. *V*, 47; *A*, VII.ii; *E*, 124).

In this way the visual signs and the things signified are complicated, twisted, knotted, or concreted together. What were merely common or shared names are now supposed to name common or shared natures. It is easy to conceive how this occurs. Consider that the same consequence often ensues with languages that are neither universal nor learned in infancy. This popular supposition, that of Direct Realism, suits well enough with the ordinary purposes of life, and probably would not have been questioned by optical theorists were it not for the problems connected with illusions. If Rule VIII refers to sort-crossing, then this rule refers to sort-trespassing; or, if it is allowed that Rule VIII refers to the use of metaphors, then this rule refers to their abuse.

What is the supposed identity of nature referred to in this rule? Now it is a fact that in our ordinary metalinguistic talk, that is, our talk *about* visual language, we do use many of the same words to refer either to its words or to the things they signify. We use, for example, the word "round" to refer to the shape that we see of a watch or to the shape that we feel. Similarly we use such

34. Cf. Vasco Ronchi, *Optics: The Science of Vision* (New York University Press, 1957, tr. by Edward Rosen from *L'Ottica Scienza Della Visione*, Bologna, 1955), sec. 105: "The finger or object is always the same, whether it is placed near the eye or far away. Yet what the observer sees changes perceptibly and progressively from one position to the other. How can two things be deemed identical when one of them remains constant while the other varies?"

words as "square," "big," "small," "high," and "low" to refer to the shapes, sizes, and positions that we see, or to the shapes, sizes, and positions that we feel. But in addition there is a strong temptation to think that these common words name common natures such as roundness, squareness, etc., that is, we think that shapes, sizes, and positions can be seen as well as felt.

Does this mean, then, that it must always be false to say that we see what we feel; that we see the same shapes, sizes, and positions that we feel? Similarly, must it always be false to assert that we hear a coach, or see a red-hot bar of iron? This is like asking whether it must always be false to assert that ice looks cold or water wet. More illuminatingly, it is like asking whether we speak falsely if we say that there is an iron curtain between the East and West. Probably we do so if we mean that there is actually an erection made of iron. Since we do not mean this but do mean merely that the boundary is impenetrable, our assertion can be true. Similarly we need not speak falsely when we say that we hear bitter words or a big noise or a high note. But we can speak falsely if we think that these "flavors," "sizes," and "heights" can be heard in the same way as we hear sounds. Similarly we need not speak falsely when we say that we see a red-hot bar of iron or a round watch.

3. The Visual and Tactual Square

In the preceding account of the linguistic solution, it was as though I started by imagining the predicament of the people in H. G. Wells' story *The Country of the Blind*. These people, completely ignorant of the visible world, understood the concept of physical object. They had words to describe the sizes, shapes, positions, and distances of objects as well as words referring to the objects of all the senses except sight. Then it was as though these people were suddenly made to see. According to my theory, confirmed by numerous experiments, they were at first unable to see things in space. Colors of various hues and intensities had no meaning for them at first sight just as the sounds of a strange tongue have no meaning for a foreigner at first hearing. Then it was as though I made the problem confronting these people the main problem to be solved. If colors and the sizes, shapes, and positions of objects are as unlike each other as the sounds and meanings of an ordinary language, how did these people manage to confound them? According to my solution, confirmed by tests on people learning to see, they did manage to confound them to an extraordinary degree. In a while they probably attained to that degree of fusion to which most of us have long since attained. The solution to this problem of vision was provided in all its essentials. Painstakingly, by the slow steps of experience involving countless ostensive definitions of the visual data, involving memory (remembering contexts and remembering previous responses to similar stimuli), involving imagination, and involving the speed that the use of a metalanguage gives to learning a language, associations were established that enabled these people to see things in space. These were some of the ingredients of my solution. In its early stages it was exactly like the solution to the problem of learning a new language.

Nevertheless the solution that I gave in the last three rules may be objected to. It may be granted that there is not enough resemblance between the "heights" of sounds and walls, between the "depths" of sounds and wells, between the "sizes" of bells and noises, and between the "wetness" and the "coldness" of the looks and feels of water and ice to justify my saying that these common words refer to such common natures or common characteristics as *height, depth, size, wetness,* and *coldness.*

But it may not be granted that the shapes, sizes, and positions of objects that we can see and feel are of the same order. Here there is identity of nature or, at least, very close resemblance. Consider the shapes of two cards before me, one square, the other round. We use the same word "square" to refer to the way the former looks or to the way it feels. Are they not examples of the same squareness? Moreover, the objection continues, the visual square is much more like the tactual square than the tactual circle. This is because the former pair have four corners and four sides in common, while the tactual circle has none. If this is so, then, contrary to my solution, we see and feel the same or, at least, closely similar shapes.

In order to answer this objection, consider the case of a blind Englishman who sits before these two cards, one square, the other round. With cards such as these so many of the tests upon once-blind people have been performed. He can easily recognize the square and the circle by touch because he already knows what they are, and he has long since given them the names and the numbers that others do. After the operation to restore his sight has been performed and after the surgeon is confident that he has attained visual acuity, he is asked whether he can recognize the square and the circle by sight alone. This, according to the reports, he cannot do. My Rules I to III represent his predicament at this stage.

After several weeks or months of experience of touching while he sees and of learning where to twist his head and how to turn his eyes, he learns to define ostensively the visual circle (i.e. certain closely similar visual objects). To cut a long story short, eventually he shows by his actions or words that he can correctly interpret these visual objects; that is, he can recognize the square and the circle from what he sees, and he can be said to see them. My Rule IV represents his achievement at this stage.

But this once-blind Englishman is no nonconformist. He has long since adopted many of the conventions of his sighted teachers. He too wants to talk in English about this brave new world of visual objects; and when he does so, he adopts the convention of calling them by some of the same names as those of the things they now so readily suggest to him. Thus he uses the same words "square" and "round" to refer either to the way the corresponding card looks or to the way it feels. But he knew long before his operation that a square must have four corners and a circle none. Accordingly, he must call the visual square "four-cornered": he contradicts himself if he calls it "three-cornered" or "round." This convention is represented by my Rule VII.

In order for this to occur, however, there need be no resemblance between what he sees and what he feels. The colors he sees when he looks at the square card might have been ostensively indicated as the round one or as a kangaroo. It just happens to be the case, or, if we extend the language metaphor the whole way, is arbitrarily ordained, that these colors indicate the square card. The reports show that once-blind people find no resemblance whatever between the shapes of the cards they feel and the new visual objects. Visual squares and triangles, for example, are called round.[35] Doubtless if visual circles were ostensively defined as squares the once-blind would find four corners in them. Numbering, like naming, is at first arbitrary and might have been otherwise. But once the correlation has been decided upon it ceases to be. The linguistic model illustrates this. While the choice of certain marks to represent certain sounds was arbitrary or might have been otherwise, their continued use is not. Once we have learned and accepted the written and phonetic alphabets, we represent the sounds skwɛ:ʀ by the marks SQUARE or by others more or less like them. If we use instead the marks CIRCLE, we are either contradicting ourselves or being unconventional. But there is no resemblance between these sounds and marks. This stage is represented by my Rule VIII (Cf. *E*, 143).

Once this man uses the same words such as "shape," "square," and "four-cornered" to refer to what he sees as well as to what he feels, he is strongly tempted to think that these common words refer to common characteristics. This temptation is reinforced by his teachers who have never questioned this supposition; and he readily succumbs. He now thinks, as they have always done, that the square shape of the card can be seen as well as felt. Can he not count the four corners by looking as well as by feeling? Since he can now see and feel the same square card, the square he sees must be four-cornered. But we now see that if we ask the question "Is the visual square more like the tactual square than is the visual circle?", we might just as well ask "Are the marks SQUARE more like the sounds skwɛ:ʀ than are the marks CIRCLE?" The supposition described here is represented by my Rule IX.

4. Learning to See Situation, Size, and Distance

The main features of my solution are apparent from the preceding account, but by extending the metaphor of ordinary language I now offer some important details. The main problem in the philosophy of ordinary language, that of bridging the gap between words and the things they signify, is duplicated in the science of

---

35. Cf. Home's case, 1806, quoted in von Senden, *Space and Sight*, p. 107: "A square blue card, nearly the same size, being put before him, he said it was blue and round. A triangular piece he also called round."

vision. If the connection is not innate, it must be learned like any other skill.

We get skill in seeing as we get skill in understanding words, and in both skills the imagination has a fatter part to play than the signs themselves. Certain sounds make me think of the trees in the park because these sounds were frequently ostensively defined for me in childhood. When they are uttered I often get images which I refer to the trees in the park, but I do not now have to go to the park to understand these sounds. Similarly, certain visual data make me think of (enable me to see) the trees in the park because certain correlations were established in my infancy. When I now *see* these visual data I refer them to the trees in the park. But I do not now have to walk to the park to "understand" them. Just as words, like counters in gambling, although "cashable" in things, are rarely "cashed," so we rarely need to confirm visual guesses, although, if there is any doubt (such as we may feel while looking at certain textures, or as Macbeth felt about the dagger that he saw before him), we try to do so.

But can my solution account for those cases of seeing where the visual words have never been cashed and probably never will be? We can see the planet Mars, the moon, and the Matterhorn although we have not been to these places. On the now commonly received Nativist or Common Sense theory, according to which things are, as it were, seen "straight off," these facts present no difficulty. They present none for mine either. There are many words, like "boomerang," "Peking," and "Sir Winston Churchill," that we understand perfectly well although they have never been ostensively defined for us. We understand them because things like them have been ostensively defined for us. Similarly we have had direct contact with small discs and balloons, and we have all climbed small Matterhorns.

We are often at a loss to understand words either in isolation or out of context. Questions like: What is time? and What is freedom? are meaningless, but many sentences containing "time" and "freedom" are understood. This fact about language is of enormous help in illustrating vision. If a bright light is detached from its usual context or is suspended in isolation, no one looking at it can be sure of its distance. But of two visual objects, equally large, the one that is fainter and higher up can suggest a far greater size than the other. Again, if one of these objects is blurred, it will seem nearer than the other. Only if the visual data are presented in their usual setting with their usual accompaniments are we confident about what we see.

Just as a large part of learning to understand words consists in learning how to respond to them, so is it the case in learning to see. To see an object is to recognize it, to know roughly how far away it is, its size, shape, and position in relation to us. But this involves knowing how to do things with it, remembering how long it takes to reach it, and feeling incipient movements in our muscles. Only a fragment of all that is fused into the object seen is actually presented. The rest are added at the instant we get the visual cue. What are added are remembered associations including remembering how we acted in similar cases. We need no optician's help to tell us whether our children can see things in space. We notice whether they can move their heads and eyes appropriately and move about successfully. Just as we can learn a language by practice or "know-how" without knowing its rules of grammar, so we can be skilled in seeing without knowing the laws of geometrical optics. None of us, when we see an object draw near or away from us, achieves this by doing geometry.

It is true, however, that we use all the cues we can get, in order to know how to act about things and thus prevent ourselves from being taken in by the so-called visual illusions. The more cues we get, the less likely are we to be deceived. The illusions of vision may be actually *seen* but they are delusions only if we misinterpret the signs. It is probably the case that tactual illusions, although not nonexistent, are very rare. It is hard to imagine Macbeth saying: "Is this a dagger that I feel within my hand, its handle before mine eyes? Come, let me see thee. . . . I see thee not, and yet I feel thee still." From the fact that we constantly use these cues in order to understand words, I infer that we use them in order to see. The context theory of meaning applies to both languages. Moreover, just as we tend to overlook the cues we use in one language, so we tend to do so in the other. Keen awareness and skill in introspection are needed in both cases to find them. Thus in addition to visual signs there are items furnished by the other senses that accompany vision. The main ones, however, come from sight and touch: not only the various characteristics of colors but also the great variety of items that come under the heading of muscular sensations. Many are discovered by using J. S. Mill's canons of inductive science, notably the methods of agreement and difference.

Let me apply the linguistic model to the specific problem of how we see the situations of objects. I know that the words HIGH and LOW, UP and DOWN, LEFT and RIGHT, etc., do not have to be high and low, etc., in order to mean *high* and *low,* etc. I know that the marks ERECT and INVERTED do not have to be erect or inverted in order to mean what they do until it is ordained what their situation will henceforth be among themselves. It would be queer to expect these situation words to have a situation with respect to what they mean, but not queer to expect them to have a situation among themselves. Yet I also know that the words HIGH and LOW, etc., share the same names as *high* and *low,* etc., that they mean. Finally, I know that there is no difficulty in bridging the gap between these situation words and the situations they are used to mean. Grammarians and children bridge it, not by using grammar rules or by inference but merely by habit. To either, the word LOW suggests *low* because it has been learned ostensively, as it would have been equally well if the custom had been to write it WOL and to pronounce it **haγ**.

Using these features, I know that visual data do not have to be high or low in order to suggest heights or depths to us; that they have a situation among themselves of a different order from the situations of the things we perceive, even though these situations are called by the same names; that there is no difficulty in bridging the gap between the different situations of visual objects and the things they suggest to us; that optical theorists and children bridge it, not by using the laws of optics or by any kind of inference but merely by habit; that the visual objects that now suggest kangaroos to us might have gone together with wallaroos; in which case they might have suggested the latter to us just as well as they now suggest the former; and that not only do the colors which suggest situations to us get the names HIGH, LOW, etc., of their referents but, from the rules previously adduced that govern other languages, additional properties; specifically, we confound the one with the other believing that we *see* and perceive by sight the same high or low things.

Using the linguistic model I know that there are additional cues, that there are items that attend the act of vision, and that they are noticeable although rarely noticed. The people in the country of the blind, before they were given sight, related the situations of objects to their bodies just as we do: *up* and *down* in relation to where the head and feet are felt to be, and *to the left* and *to the right* in relation to where the left and right hands are felt to be. On being made to *see,* their field of vision had no objective locality whatsoever. Slowly they learned to see, and this included confounding the attributes and appellations of the things thought of with the visual data. The cues they used included minute sensations that accompany the turn of the eyes and the massive sensations that accompany the movements of the head.

In using the linguistic model to illustrate how we see size, similar considerations apply. Just as I know that the marks BIG and SMALL do not have to be big or small in order to mean big and small; that others might have been used; that though the marks ORNITHORHYNCHUS are bigger than PARIS, they suggest something a hundred thousand times smaller; but that the size of the word has nothing to do with this suggestion; that words have a size among themselves; and so on; so I know similar things about the visual data and their relations to what they signify. I can infer that visual objects are large or small, but that their size is of such a different order from the size we think of that it should, perhaps, be given a different name; that of two visual objects equally large, one may suggest a hundred thousand times greater size than the other, for example, the moon and sixpence; that a man fifty or five feet away "looks" equally big; that whenever we ordinarily say that a house is big and a cup small this must be meant of the house and cup we think of; that the visual size, although noticeable, is usually overlooked; that the visual size alone is useless for perceiving the size of objects; and that other signs or cues having little or no resemblance to what they signify are needed to enable us to *see that* physical objects have a certain size. I can know what some of these are from the linguistic model, for example, those furnished from memory and imagination and from the contexts in which visual objects appear. These turn out to be the most useful ones. But from the items that accompany the act of seeing, I can infer that we use such cues as the faintness and clarity and the confusion and distinctness of the visual object, as well as the sensations that accompany the turn of the eyes inward to see things close up, and the feeling of strain that we experience when we try to prevent the blurriness, fuzziness, or confusion of the visual object. For example, disregarding other cues, a blurred visual object suggests smaller size than a distinct one, while a faint one suggests greater size than a clear, vigorous, or intense one.

Similar considerations apply to how we see distance, but there are interesting differences, for the discovery of which the linguistic model gives me a lead. Just as the words NEAR and FAR do not have to be near or far from what they mean in order to mean what they do, so I know that visual objects do not have to be near or far from the objects they so readily suggest to us. Just as it is non-sense to ask how far away the marks CUP are from the cup we imagine, so it must be nonsense to ask how far away a certain combination of colors is from the thing we can point at. I conclude that visual objects are equidistant from, or, more correctly, are at no distance from the things we think of. To say that they are is either to abbreviate or to mix different categories, as if we were to say: "He came into the room smelling of musk and insolence." So far the parallel with size and situation is close. But while it is true that visual objects have a size and situation among themselves, the same cannot be said of distance. The marks CUP are erect in relation to the other marks on this page; they are to the left of, to the right of, above, and below, others; they are larger and smaller than others; but they lack the dimension of depth or distance among themselves. Like rainbows, they have no backs to them. I can infer that the same is true of visual objects. Nevertheless, like the marks on this page or their corresponding sounds, although they lack the dimension of depth in relation either to the things they suggest or to other visual data, it is clear that the visual data are outside us. That is to say, they are not in our eyes just as sounds are not in our ears. Now although they thus lack the depth or distance characteristic of the things we see, the visual data manage to suggest objects at various distances to us. They do this in much the same ways as the marks NEAR and FAR suggest near and far things to us and in the ways already indicated for vision in general.

5. Conclusion: Visual Data and Mind-Facta

In what precedes I have tried to give the main ingredients of a preliminary solution to the problem of vision. I tried to show how to bridge the gap between what is sensed and what is perceived. The bridge I used was a principle that Aristotle only hinted at but perhaps used more than he said, namely, that the visual data form the elements of a language. In this enterprise I simplified my task enormously by providing an indefinite interpretation of the signs of visual language and a narrow interpretation of the things signified. I interpreted the visual signs merely as colors of various hues and intensities variously combined and ordered; and I conceived the things signified as any of the spatial properties of physical objects. Because I assumed that a blind man who has never seen could know all about them, I identified them with tactual objects. Equipped with these limited interpretations, with the help of some of the features of the linguistic model I showed how we see. Although this was a convenient simplification, the main elements of the whole solution were present. My aim was to bridge the gap by treating tactual objects as if they were physical objects. But such interpretations are too confining. Visual signs and the things signified are much richer than I prescribed. These wider interpretations I began to insinuate as my account proceeded.

It will be seen from the last section that much more can be said about the nature of the visual data than may at first be realized. Their features, although noticeable, are usually overlooked. Rule VII shows that when we talk about visual data we are talking about the elements of another language. Once we realize this we begin to dispel much of the obscurity that surrounds their nature. For we see at once that we talk about these elements in at least two important ways: (1) in the idioms appropriate to themselves, and (2) in the idioms appropriate to their referents. Rule VIII shows that when we speak in the latter way we cross sorts whose members have no more resemblance than do cats and "cats." These two ways may be characterized as speaking literally and speaking metaphorically; or, more appropriately, because the metaphors are partly submerged, as using a primary and a secondary vocabulary. In the primary vocabulary we can talk of colors of various hues, such as *blue, green, red,* and so on; and these may be *confused* (or *fuzzy* or *blurred*) or *distinct,* or they may be *faint* (or *dim*) or *clear* (or *intense*). Such is the scantiness of our ordinary language, however, that in most of our talk about the visual signs we must borrow words from the vocabulary of the things signified. These words constitute the secondary vocabulary. We need to refer to those features of the visual signs that we call their *sizes, shapes,* and *positions.* We need to say they are *large* or *small, round* or *square, high* or *low,* that they are at a certain *distance* from us, and so on. But because we use these words which are really metaphors it cannot be too strongly stressed that there is no more resemblance between the size, shape, position, and externality of a visual datum and the thing we see than there is between a big or small noise uttered in a high or low key and the thing it is used to mean.

Most of the time, however, we are talking not about the visual data but about our interpretations of them. We say that a man *looks* just as big from ten feet away as he does from five, but it is obvious that we are referring to the size we think of. Nevertheless, although I cannot now see a color without adding its meaning, I can roughly measure visual size. In my visual field my thumb is half as big as the Eiffel tower, while my hand is as big as the sky. We say that the Eiffel tower *looks* erect whether we view it standing on our head or our feet, but once more it is obvious that we are referring to the situation we are thinking of. Nevertheless the objects in my visual field have various situations among themselves. We say that the moon *looks* extremely remote, but so does the moon in Whistler's painting. In this case, however, although both visual moons are seen outside us neither is more remote or nearer than the other. Yet we see them a long way off.

From my account does it not seem that visual data are such that we cannot make mistakes about them? Are not visual data invented for the purpose of avoiding error? It is true that with the help of the linguistic model I can see at once that we cannot make mistakes about the visual data. But the price is high. If no deception is possible here neither is any non-deception. For if the visual data are merely characters on a writing tablet on which nothing as yet stands written, the question becomes: Are characters or words true or false? To try to discover the truth or falsity of visual data out of their context would be as hopeless as to try to discover the truth or falsity of words out of their context in a sentence. Visual data in themselves, therefore, are devoid of any significance. We may be acquainted with them but we cannot know them. We may see them but we cannot *see that* they are true.

I must now deepen and widen my interpretation of the things signified by the visual data. As a preliminary device, like a scaffolding, that can now be discarded, I treated sizes, shapes, positions, and distances as tactual objects, which they are not. Nevertheless we can tell by sight the sizes, shapes, etc. of objects.

What sort of things do we see? Let me begin by returning to the account I gave at the commencement of the whole problem. We commonly say and believe that we see physical objects in space, such as tables, sticks, stones, apples, bells, daggers, lakes, the Eiffel Tower, and the moon; some nearer, others farther off, of various shapes, sizes, and situations; physical objects that are colored, hot or cold, rough or smooth, and that sometimes have taste, smell, and resonance. Moreover, we say and believe that the apple we see is the same thing as the apple we feel, smell, and taste, and that we see *one* thing, the apple, whose red color, round shape, and sweet taste are qualities of *it*.

Is it not physical objects such as these (although perhaps differently described), not tactual objects, that we perceive by sight? It is not my purpose here to give a complete description of their nature. Nor do I pretend to know what it is. But the account I gave of the objects of sight and touch was a model for understanding how we see these physical objects. The rules that I used to show how visual and tactual objects are complicated or concreted together can be extended to accommodate the objects of all the senses. Just as certain visual data, having no more resemblance with the tactual square than CATS have with cats, become a sign of the tactual square, so do certain visual data become a sign of the apple we perceive by sight. Just as we learn to tell by sight the shape that I called the tactual square, so do we learn to tell by sight not only the shape, etc., of the apple but also its smell and taste. Just as we want to say that what we feel *looks* square, so we want to say that ice *looks* cold, an apple *looks* tasty, a gong *looks* noisy, a suitcase *looks* heavy, and a carcass *looks* smelly. Just as we use the same name "round" to refer to certain similar looks and to certain similar feels, so we use the same name "apple" to refer to certain similar looks, feels, tastes, and smells. Just as we are strongly tempted to think that the now common word "round" refers to a common nature or abstract idea, and that we see and feel the same round shape, so are we strongly tempted to think that the word "apple" refers to an abstract idea; that we see, feel, smell, and taste the same apple; and that *it* is a subject or substance distinct from its predicates or qualities of redness, roundness, and sweetness. But we have seen that just as we want to say that we see the same round shape that we feel, so we want to say that we see the same apple that we feel, smell, and taste. We have seen, moreover, that it would be a mistake to think that these assertions and others such as "I hear a coach" and "I see a red-hot bar of iron" must always be false.

These considerations suggest the inadequacy of interpreting the sign and the thing signified as the visual datum and the physical object, respectively, if the latter is conceived as a sort of receptacle for its qualities though distinct from them. But they also mirror a defect in the linguistic model, specifically in the feature of sign-thing signified, if the latter is narrowly conceived, for example, as noun-nominatum. Now although I intend to keep the sign-thing signified feature for general use, I need to supplement it in order to illuminate the aspect of vision I am now considering.

Consider the statements of the once-blind person. He says of the watch: "That's round; it's a watch," although he cannot yet tell by sight that it is one. Six months later he may say: "In the dark I hear my watch; I turn on the light and see it; I reach over and pick it up." Both sets of statements indicate that he possesses the concept of a physical object as a thing with its qualities or a substance with its attributes. He is making use of a remarkable conceptual device probably invented by some great sort-crosser of the very remote past for the purpose of sorting matters of fact. For it represents the facts of one sort in the idioms appropriate to another. Specifically, it represents states of affairs as if they were subjects and predicates of sentences. That it is such a device is confirmed by the fact that many of its users use the terms "subject," "substance," and "thing," on the one hand, and the terms "predicate," "attribute," and "quality," on the other hand, interchangeably. The subject-predicate device is of enormous utility. So useful is it and so common has it become that what once was sort-crossing is now sort-trespassing. What was once a lively metaphor is now a myth unconsciously believed in. For we ignore other ways of sorting the facts.

We *use* the subject-predicate device when we make believe that just as the sentence "A watch is round" has a subject and predicate, so the word "watch" refers to a subject, substance, or thing whose predicate, attribute, or quality is roundness. We *use* it also when we make believe, in accordance with the sentences "I see the watch" and "I see the shape of the watch," that we see the substance watch or that we see a quality of it. But we are *used by* it if we believe that we actually see the substance watch or the roundness of it.

Now from the conclusions reached in my discussion of metaphor we see that it would be vain to say that such assertions as "I see the watch" and "That's round; it's a watch" must always be false. We might just as well say that such assertions as "Visual data are signs of things signified" or "There is an iron curtain" must always be false. But we can say that the device does not properly illuminate. Indeed it sheds too much light, for its uncritical use strongly tempts us to think that there is more in the visual object than we actually see. Nevertheless, the subject-predicate device can be specified as a feature of the linguistic model. I can use it as an auxiliary of the device of sign and thing signified. Thus the achievement of seeing the round shape of a watch may be represented as seeing a predicate or quality which is of a subject or thing. But in turn the predicate roundness may be represented as a sign, and the subject watch may be represented as the thing signified by it. In this way the two devices could be made to mesh, signs representing the data conceived as predicates or qualities, and things signified the physical objects conceived as subjects or substances. I could let the matter rest at this stage. The subject-predicate device, however, contains ingredients that I do not want. The subject or substance *owns, has,* or *contains* its predicates or attributes. Must the thing signified *own, have,* or *contain* its signs? Auxiliary models, like other metaphors, should be consistent. To use these two together would be like saying: "The introduction of the burning question chilled our spirits." Something more is needed in order to free the concept of the physical object that we see in space from the notion that it is a receptacle of qualities.

Let me, therefore, explode the subject-predicate myth by inventing a new metaphor. In order to do this I choose another feature that I specified in the linguistic model, that of definiendum-definiens. Now when we say that a watch is round, has a face, and goes "tick-tick" etc., we can abstain from holding that we are referring to the qualities of a thing or predicates of a subject, and need only hold that we are explicating the meaning of the word "watch" by saying what other names are appropriate. I apply this device to the subject of vision by representing the facts as if they are definienda and definientes instead of subjects and predicates. Thus the qualities of the watch turn out to be nothing but definientes and the physical object becomes the definiendum.

So now when we say "I see a watch" or "I see the color, size, shape, etc., of the watch," we can abstain from thinking either that we have achieved the feat of seeing the substance watch itself or that we have located by sight the proper receptacle for these attributes of color, size, shape, etc. We need only hold that we have succeeded in finding the meaning of certain visual words: these visual words we call in our metalanguage "white," "small," "round," etc., and their meaning, when they are thus conjoined in our experience, we call "watch." Thus the physical objects we see in space are meanings that we assign to visual words. But if these visual words that constitute some of the definientes are only characters on a writing tablet on which as yet nothing actually stands written, then their meaning that constitutes the definiendum is appropriately expressed in a sentence that can be true or false. Accordingly, the statement "I see a watch" abbreviates "I *see that* it is a watch" which, in turn, abbreviates "I *see that* this and that and that, which have the right to be called 'white,' 'small,' and 'round,' also have the right, when taken together, to be called 'watch.' " Moreover, in accordance with my interpretation of Aristotle, if the visual data are only words, then their meanings, the physical objects, are mind-facta. This shows that the simple statement "I see a watch" records a complex conceptual act;[36] it shows that the verb "to see" is certainly an achievement word; and it shows that when we see physical objects we are makers or poets.[37]

Which, then, is the better make-believe, that the visual data are predicates of a subject or that they are definientes that mean a definiendum? These two antithetical metaphors can issue in two opposing metaphysics of vision if, instead of making believe, we believe. Let me confine the choice to one between metaphors. In choosing between rival metaphors there are tests in general that we can appeal to. These I shall adduce at a later stage. It is enough if I say here first, that if economy is better than extravagance, then the choice is easy, for it is surely extravagant to postulate as the object of sight the invisible subject distinct from its predicates; and secondly, that the choice of the definiens-definiendum metaphor or of one closely like it is the outcome of the argument I have been following ever since I was persuaded to answer "No" to Molyneux's question.

It might be objected that language is not an appropriate model to illustrate how we see because this visual language lacks a defining feature of language; it is a constant sermon or lecture whose audience is mute. Now it is true that few of us are painters, but we are certainly not dumb. On the contrary we are chattering constantly about visual language in another language, namely, our own native but artificial tongue. To see does include listening

36. Cf. von Senden, p. 298: "When I say 'I see a window,' these words are a false account of what I literally see. For 'window' is not a visual term but a conceptual one. I see, not a 'window,' but something colored, whose specific quality I know to have the meaning 'window.' "

37. Cf. Plato: "You are aware that *poiésis* is a term of broad application. In all cases where anything whatever passes from not-being into being, the cause is *poiésis;* so that [in a sense] the works produced by all the arts are acts of *poiésis,* and the makers of them are poets." Diotima's statement in *Symposium,* 205 B.C., tr. by Philip Wheelwright in his *Aristotle* (The Odyssey Press, New York, 1951), p. xxvii.

to a language, that is to say, this notion illustrates part of the nature of vision. Often we have only to listen to the lecture and the meanings appear to come flowing in upon us. We see things straight off and are not mistaken. Moreover, in the sermon of visual language, if we are not taken in by its ambiguities, we are constantly admonished by means of its hypothetical imperatives what to shun and what to approach. But vision is much more than mere listening and receiving instruction. It is a dialectic or a game of question and answer that we play with visual language. Most of our metalinguistic answering back is in the form of questions, often silent questions of the form: "What does this *mean* to me?" "Is this to be called 'a lake' or 'a mirage'?" "Is this to be called 'a crooked stick' or 'a straight stick'?" and "Is this to be called 'an army tank' or 'a fly on the window pane'?" The logic of this game, invented by Socrates, is the same as the logic of the method of scientific discovery. It involves the invention of hypotheses, the search for counter examples, then their rejection, revision, or retention. This is what happens when we are confronted with the ambiguities of visual language known as visual illusions.

In illusion, as its name implies, we are played against or mocked. At times, being genuinely cheated or deceived, we lose the game; at others, having seen through the deception, we win. It is as if we are mocked by the fiend that "lies like truth" whose discourse has two meanings, one open and apparent, the other real and hidden. Illusions are like metaphors. There is nothing in the mirror-image itself that says that it is merely a mirror image; nothing in the apparently crooked stick that says that it is really straight. Mocked by the ambiguities of visual language it is no wonder that the once-blind man asked which was the lying sense, feeling or seeing. But if illusions are games played against us we too can be players in these games. Accordingly, if Socrates defined thinking as "simply talking to oneself, asking questions and answering them, and saying Yes or No," and judgment as "doubt being over, the two voices affirm the same thing," [38] that is to say, thinking is a discourse and judgment a statement pronounced, then I should define seeing as talking to oneself and pronouncing

38. *Theaetetus*, 190.

statements about another discourse that we hear with our eyes; that is, seeing is a soliloquy conducted in one language about another while we hear it.

Let me illustrate with the dagger experiment performed by Macbeth. Thanks to his prenotions he already has a hypothesis, and he asks a question about the meanings of the words of visual language: "Is this a dagger which I see before me, its handle before my hand?" or, in other words: "What does this color combination mean? Does it mean what we conventionally call 'a dagger'?" If it does then certain results follow, for example, that it is accompanied by a well-remembered feeling in his hands. He tests his hypothesis: "Come, let me clutch thee"; and disconfirms it: "I have thee not, and yet I see thee still." Therefore, this color combination does not mean what is called "dagger." He invents a new hypothesis: It means "a dagger of the mind." Unsatisfied, he tests once more by feeling and by regarding an ordinary dagger. He asks which is the lying sense, sight or touch. Conviction that it is the former comes when the visual objects change color. These non-denoting visual expressions are correctly interpreted to mean "an hallucination." Nevertheless, they direct his action, marshaling him on the way to his fate.

In order to show further that the linguistic model is eminently suited to illustrate the facts of vision, I shall test the Linguistic Theory of Vision against a rival theory based upon the geometrical model. But first I shall select features from the Geometrical Theory that show how, starting from the Greeks who invented it, it was tortuously improved upon up to the time of Kepler, the father of modern optics.

PART THREE

TESTING THE METAPHOR

CHAPTER VI

The History of Vision

1. Seventeenth-Century Theories

EVERYONE knows that optics is the study of light. Although the name "optics" originally meant the science of sight, the light usually referred to in optics is not an object of sight. It is not light as we ordinarily use the word, and its opposite is not dark either. If our eyes are open in its presence, however, we say that there is light; and if our eyes are open in its absence we say that it is dark. It is assumed in physics that matter, consisting of small corpuscles, emits radiations of energy. The sum of these radiations is the electro-magnetic spectrum. A fragment of this spectrum is light or the visible spectrum. This light affects photoelectric cells, photographic emulsions, and eyes. Optics is thus a branch of physics, and it includes photoelectricity and photography. As its study is the nature of light itself, its propagation by waves or photons and its absorption, it is called physical optics. It has nothing to do with vision. As we have seen, it keeps the name "optics" only because its subject is light, and its subject keeps the name "light" only because it is that part of radiant energy that affects eyes. The name "photics" would be more appropriate. The names "microscopic optics" and "quantum or atomic optics" are sometimes used depending on whether the classical wave theory or the photon theory is assumed.

In so far as the purpose of optics since the time of Kepler has been the making of optical instruments, optics is called "geometrical" or "ray" or "macroscopic," because for this purpose the assumption of the ray model of light usually works. When it does not, the wave model is assumed. Euclidean geometry and the basic laws of reflection (stated by Euclid in the fourth century B.C.) and refraction (discovered by Descartes or Snel in the seventeenth century) are applied to the formulation of rules for determining the position, size, and shape of the optical image whether real or virtual. Vision enters into this science only in that, first, the image can be either projected on a screen or seen through an optical system; secondly, the eye is an example of an optical system; and thirdly, the eye is the beneficiary of the ultimate design.

This roughly classifies the main sciences now called optical, but the catalogs of some universities, schools of optometry, and institutes of optics list a fairly recent addition. This is physiological optics described as the physics, physiology, and psychology of vision. Helmholtz in his handbook on this science stressed physiology, but he used geometrical optics to account for the dioptrics of the eye, and he devoted the last of this three-volume work to the psychology of vision.

Ancient optics, however, was not thus classified. Almost all the Greek philosophers of note were interested in vision, and they used all their knowledge—which we should now distinguish as physical, anatomical, physiological, geometrical, and psychological—to describe the subject. In general, from Aristotle to Kepler, there were four overlapping problems: (1) to provide rules for determining the place and size of the object seen in direct vision; (2) to provide rules for determining the place and size of the object seen through mirrors—and, though later, lenses; (3) to describe the anatomy of the eye; and (4) to describe the mechanism of vision in the eye. These problems, it seems, were understood by Aristotle and his contemporary Euclid. The search for their solutions, continuing for two thousand years, ended, except for refinements, with the solutions given by Kepler in his *Supplement to Witelo* (1604) and *Dioptrics* (1611). Standing on the shoulders of Euclid, Ptolemy, Galen, Ibn al-Haitham, Ibn Rushd, Witelo, Bacon, Maurolico, Porta, and many others, he was able to provide solutions which together constitute the foundations of modern optics. Within a century his solutions acquired the status of dogma. Most of them were incorporated in the axioms of Newton's *Opticks* (1704).

I shall set down Kepler's solutions as five interconnected rules, and, immediately following, I shall indicate variations and enlargements upon them made by some of Kepler's followers including Descartes, Malebranche, and Newton. Although Kepler stated all these rules, in some of them I summarize what he meant rather than state what he actually said, using in many cases more familiar terms. Moreover, most of the diagrams, indicating somewhat

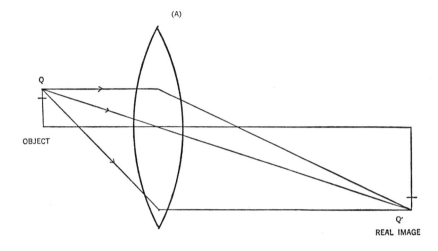

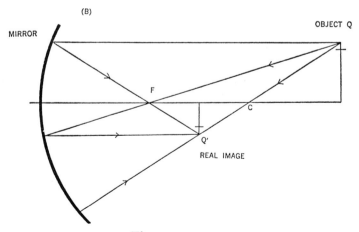

Figure 1

neater constructions than Kepler's, are based on those found in any recent text book of geometrical optics.

(1) Rays of light diverging from all points of an object and made to converge by an optical system will make an inverted picture —now called the *real image*—of the object on a screen (Fig. 1).

This is equivalent to Newton's Axiom VII. The words "diverge," "converge" (both referring to the paths taken by the rays), and "picture" were invented by Kepler and adopted by Newton. Real images are said to be produced by converging lenses and mirrors when the object is outside the focal point. They are not produced by plane surfaces.

(2) The eye is an optical system with a lens and screen on which an inverted picture, now called the *retinal image,* is painted (Fig. 2).

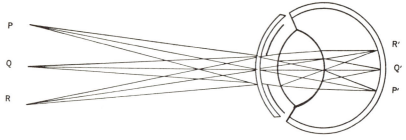

Figure 2

This rule is incorporated in Newton's Axiom VII as an illustration of it. It purports to describe the cause of vision, succinctly put by Kepler: "Sight is the stimulation of the retina," and enlarged upon by Newton: "For anatomists, when they have taken off from the bottom of the eye that outward and most thick coat called the *dura mater,* can then see through the thinner coats the pictures of objects lively painted thereon. And these pictures, propagated by motion along the fibres of the optic nerves into the brain, are the cause of vision." Kepler, being primarily a theorist, did not perform this experiment, but Scheiner and, later, Descartes were probably the first to do so. The latter, in his *Dioptrics* (1637), gave a full account from which I extract portions to show first, how Descartes thought what was later called "accommodation" was achieved, and secondly, how he tried to avoid the ghost-in-the-machine fallacy often attributed to him: "Take the eye of a newly dead man. . . . You will see (I dare say with surprise and pleasure) a picture representing in natural perspective all the objects outside . . . ; if you squeeze it never so little more or less than you ought, the picture becomes less distinct. And it should be noticed that the eye must be squeezed a little more—made proportionately a little longer—when the objects are very near than when they are further away. . . . And when it is thus transmitted to the inside of our head, the picture still retains some degree of its resemblance to the objects from which it originates. But we must not think that it is by means of this resemblance that the picture makes us aware of the objects—as though we had another pair of eyes to see it, inside our brain; I have several times made this point; rather we must hold that the movements by which the image is formed act directly on our soul *qua* united to the body, and are ordained by nature to give it such sensations." [1]

(3) "In vision with two eyes, in order to judge visible distance, we make use of the interval between the two eyes provided that the distance bears some sensible proportion to it . . . for, given two angles of a triangle with their connecting side, the remaining sides are also given." [2]

This process amounts to what is now called *convergence*. Descartes and Malebranche gave similar accounts of this rule. The former said that "we know distance by the mutual relation of the two eyes" just as a blind man holding two crossed sticks can tell, "as it were by natural geometry *(ex geometria quadam omnibus innata)*" where they intersect, given merely the distance between his

---

1 *Dioptrics*, V and VI from *Descartes: Philosophical Writings*, translated by E. Anscombe and P. T. Geach (Edinburgh, Nelson, 1954). In this chapter unless otherwise noted, all quotations from Descartes are from his *Dioptrics*, VI.

2. *Supplement to Witelo*, III. 8, my translation. The Latin is: "Cum sint singulis animantibus a natura dati bini oculi, quos inter est aliqua distantia, hoc adminiculo sensus visus rectissime utitur ad iudicandas visibilium distantias, dummodo senibilem habeat distantia illa proportionem ad distantiam oculorum. . . . Datis enim duobus angulis trianguli, cum interiecto latere, dantur latera reliqua."

hands and the size of the base angles. "The act of consciousness involved is a simple act of imagination; but it contains implicitly a reckoning like that made by surveyors who measure inaccessible places by means of two different observation posts." To this Malebranche in his *Recherche* (1694)[3] added the corresponding means for judging distance: "The disposition of the eyes which accompanies the angle made by the visual rays."

(4) "In vision with one eye we are able to use the distance-measuring triangle (*triangulum distantiae mensorium*) which has its vertex in the point of the object and its base in the width of the pupil,"[4] i.e., we judge where the object is from the divergence of the rays, computing from the retinal image—a process amounting to what is now called *accommodation* (FIG. 3).

Figure 3

Descartes made a variation upon this rule. The means used here, he said, is "the shape of the eye . . . and when we adjust it according to the distance of the objects, we also produce a change in a certain part of our brain. . . . Ordinarily this happens without our attending to it; just as, when we squeeze a body in our hand, we adjust our hand to the size and shape of the body, and thus feel the body without having to be conscious of these movements of the hand." But Malebranche went even further and explicitly interpreted the rule psychologically, making the means clearly perceptible. The medium used is "the disposition of the muscles which constringe our eyes in order to make them somewhat longer: and this disposition is, moreover, painful."[5] Ignoring these variations for the time being, what Kepler intended was first, that the divergence of the rays from a point source is the cause of

3. *Recherche de la Vérité* (Oxford, 1694, T. Taylor's translation), 1.9.3.

4. *Supplement to Witelo*, III. 9, my translation. The Latin is: "Id vero triangulum distantiae mensorium etiam in uno oculo potest considerari, ut vertex sit in puncto rei visae, basis in latitudine pupillae, et diametro pupillae ea quae coincidit cum linea connectente puncta utriusque pupillae."

5. *Recherche*, 1. 9. 3.

its appearance there, and secondly, that we solve its location by geometry.

Kepler then applied his distance-measuring triangle, thus used in direct vision, to vision in lenses and mirrors:

(5) An object seen by reflection or refraction always appears in that place from which the rays appear to diverge in falling upon the eye. This appearance of the object (or the place where it appears) is now called the *virtual image* (FIG. 4).

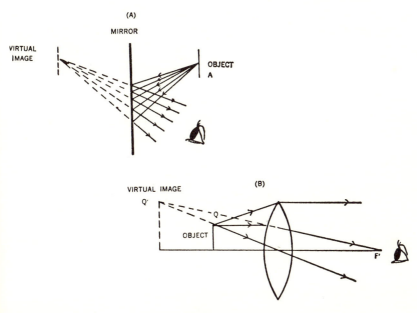

Figure 4

Kepler used the names "picture" and "image" for the real image and virtual image, respectively. The latter is called "virtual" because the rays do not actually pass through it; they must be projected to find it. Virtual images are said to be produced by converging lenses and mirrors when the object is inside the focal point, and by diverging lenses and plane and convex mirrors when the object is in any position behind the lens or in front of the mirror. This rule is Newton's Axiom VIII, and the diagram (FIG. 4A) copies his. He gave the reason why the image is so placed in terms of Rules (2) and (4): "For these rays do make the same picture in the bottom of the eyes as if they had come from the object really placed at A without the interposition of the looking-glass; and all vision is made according to the place and shape of that picture," which shows that, although he did not actually state it, he took Rule (4) for granted.

In this fashion Kepler solved the problems set by the Greeks, and he did this by continuing to use their greatest invention, the geometrical way. The pattern of modern optics had been cut, and so it seems, the problem of vision solved. We all have a built-in geometry which we use, more or less without taking notice of it, to see objects in space, just as a surveyor locates objects by triangulation if we suppose him unheedful of the process. In this presentation of Kepler's contribution I have omitted to stress certain features: his treatment of the pupil as a diaphragm; following Ibn al-Haitham, his idea that rays emanate from every point of the body; and his concept of rays emitted in a cone from the point source, of which his triangles are convenient cross-sections.[6] I have also omitted to stress that his achievement, now commonplace to all but the historian who knows what went before, was the work of genius.

2. Classical Optics

These five rules are the basis of a geometrical theory of vision. I have already noted how some psychological elements were fused with it, and shall note more later. However, owing to the Greeks, ancient or classical optics was also predominantly geometrical. As we have seen, the Greeks posed the main problems. Their own solutions correspond to Kepler's five rules, and the great optical theorist whose solutions remained as received theory for two thousand years, and in part still remain, was Euclid. He realized along with Aristotle that optics fell under its superior science geometry. And so his *Optics* and *Catoptrics,* the earliest extant works on geometrical optics, are applications of the geometrical model to vision. We can guess without reading these works that the points, lines, and angles of his *Elements* will reappear. It was a momentous achievement to interpret lines as rays of light and to make

6. For details of these lucidly presented, see Vasco Ronchi, *Optics: The Science of Vision,* Ch. II.

them obey the rules of geometry. Keeping in mind Kepler's five solutions to the main problems I shall present the corresponding solutions found in classical optics.

It should be noted, however, that from the standpoint of geometry these solutions are independent of a controversy that lasted throughout the whole classical period and that was not decisively settled until Kepler's time. This is the controversy between the emanation or emission theory of vision and the immission theory. According to the former, held by the Pythagoreans, Plato, Euclid, Galen, Ptolemy, Roger Bacon, and Grosseteste, visual rays emanate from the eye to clothe the object. According to the latter, held by the Atomists, Aristotle, and Ibn al-Haitham, visual species or facsimiles of the object or, eventually, rays of light, enter the eye. But in geometrical constructions the lines are the same whether drawn to or from a point.

Kepler's Rules (3) and (4) describe the geometry of direct vision. What did the Ancients have in their place? Corresponding to Kepler's Rule (4) which describes how we see with one eye by using the distance-measuring triangle, are two of Euclid's postulates from his *Optics,* called "definitions" but preceded by "let it be assumed." Just like Kepler's triangle, which is really a cone, here we find another triangle, only this time the direction is reversed:

> The form of the space included within our vision is a cone with its apex in the eye and its base at the limits of our vision. Those things seen within a larger angle appear larger, and those seen within a smaller angle, smaller; and those things seen within equal angles, appear to be of the same size.[7]

This rule withstood the test of centuries. Roger Bacon's adoption of it is representative, and the diagram (FIG. 5) copies his: "Vision takes place along the cone (*pyramis*) whose vertex is in the eye and base in the thing seen. . . . The same object at a distance makes a small angle, which when near would make a large one." [8]

To Kepler's account of convergence—his Rule (3)—the solution

7. "The Optics of Euclid," *Journal of the Optical Society of America,* Vol. 35, No. 5 (May 1945), tr. H. E. Burton.

8. *Opus Maius,* tr. R. B. Burke (Philadelphia, 1928), pp. 471, 523.

of classical optics, if we ignore the question of emission or immission, not only corresponds; it is the same. Euclid, to my knowledge, did not mention it, but Roger Bacon did: "In vision with two eyes, the object appears where the two axes concur." [9]

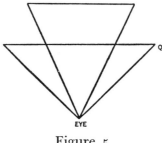

Figure 5

This completes the account of the corresponding rules for direct vision of Kepler and of the ancients if I ignore for the time being the psychological rules which were added to these. We shall see that Kepler's followers mixed psychological rules of vision with geometrical ones just as the ancients did.

We now come to the ancient solutions of the problem of vision through optical systems. What were the rules for determining the place of the object seen through mirrors and lenses? What did the ancients have corresponding to Kepler's Rules (1), (2), and (5) that show how to construct real and virtual images? One would expect nothing, at least from the Greeks, for they had no lenses, and their mirrors must have been crude. Yet, right at the start, they were able to discover the law of reflection and a rule for locating images that still survives in that it has been absorbed by Kepler's rules. Thus corresponding to Kepler's Rule (5) for determining the place of the virtual image, were Theorems 16, 17, and 18, of Euclid's *Catoptrics*:[10]

In any plane, convex, and concave mirror, the object appears at the junction of the visual ray and either the perpendicular drawn from the object to the surface of the plane mirror or

9. *Opus Maius*, p. 511.
10. Euclide, *L'Optique et la Catoptrique* (Paris, 1938, tr. Paul Ver Eecke): my translation: I have joined the statements of the three theorems.

the line drawn from the object to the center of the sphere of
the convex or concave mirror.

These three theorems, it seems to me, are not deducible from
his postulates, but the diagrams, which copy Euclid's own, show
clearly what he had in mind. The first two (FIGS. 6A and B) show

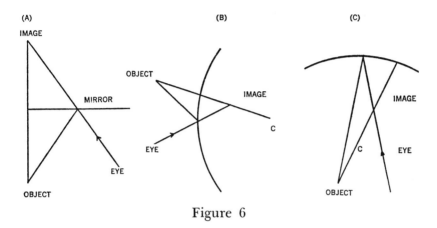

Figure 6

graphical constructions of what is now known as the virtual image,
but the third (FIG. 6C) shows how to construct a real image because, although Euclid did not say so, it is inverted in relation to
the object, and because on the emission theory (Euclid himself
was an emissionist), the visual ray actually passes through it.

Because Euclid's rule shows how to construct the real image, we
must consider it as corresponding not only to Kepler's Rule (5)
but to his Rule (1). Similar accounts were given by Ptolemy, Ibn
al-Haitham, Witelo, Grosseteste, and Roger Bacon, which show its
independence from the rival emission and immission theories. In
his *Book of Optics* (c. A.D. 1000), the Arab astronomer Ibn al-
Haitham, who relinquished the emission theory, stated the rule
somewhat differently: "In any plane, spherical convex, or spherical
concave mirror, the image is seen at the junction of its perpendicular of incidence and the reflected ray" [11]—the phrase "the perpendicular of incidence" being an immissionist's name for Euclid's

11. *Opticae Thesaurus Alhazeni Arabis* . . . (Basillae, 1572, tr. Federico Ris-
nero), Bk. 5, Ch. 2, my translation. I have joined three theorems.

line drawn from the object to the surface of the plane mirror or to the center of the convex or concave mirror.

So far we have seen the rule applied only to reflection, but it applies equally well to refraction. Euclid never made this step. He wrote a *Catoptrics,* not a *Dioptrics.* The last postulate of his *Catoptrics* shows only that he was aware that visual rays were bent when passing through water enabling objects otherwise hidden to be seen. Five centuries later, however, Ptolemy in his *Optics*[12] showed that he was probably aware that Euclid's rule for reflection applied to refraction at least at plane surfaces, because he measured the angles of refraction made by visual rays passing from air into water and glass and the reverse. In the Middle Ages the extension of the rule was well known. Grosseteste's and Roger Bacon's accounts are representative. The former reads: "A thing that is seen through the medium of several transparent bodies does not appear to be as it truly is, but appears to be at the junction between the ray passing out from the eye in continuous and direct projection, and the line coming from the thing seen which falls on the surface of the second transparent body nearer the eye at equal angles on both sides (*ad angulos aequales undique*)." [13] Bacon stated it more concisely: "Vision by refraction is at the intersection of the visual ray with the cathetus, as has been stated in regard to reflection," [14] "the cathetus" being a neutral name for the perpendicular of incidence. He used the principle to explain why a straight stick looks bent when partially under water. The diagrams, which modify or copy Bacon's, show constructions of a virtual (FIG. 7A) and a real image (FIG. 7B) after refraction at a plane and spherical surface respectively.

We now see that Euclid's rule when applied to refraction as well as reflection is the classical counterpart of Kepler's Rules (1) and

12. *L'Ottica* (Torino, 1885), pp. 144 ff. Cf. Ronchi, *Histoire de la Lumière* (Paris, 1956), p. 21.

13. *De Iride* from L. Baur, *Die Philosophischen Werke des Robert Grosseteste, Bischofs von Lincoln* (Münster, 1912), pp. 74–75, translated by A. C. Crombie in his *Robert Grosseteste and the Origins of Experimental Science,* p. 123. The last phrase enables Dr. Crombie to give a different interpretation of the passage, but it is pretty certain that Grosseteste intended the line to be the perpendicular of incidence. See Colin M. Turbayne, "Grosseteste and an Ancient Optical Principle," *Isis,* Vol. 50, Part 4, No. 162 (December 1959), pp. 467–72.

14. *Opus Maius,* p. 565.

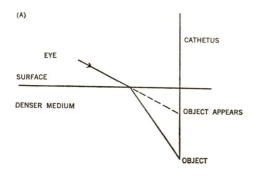

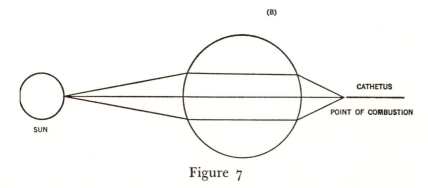

Figure 7

(5). Isaac Barrow, in his *Eighteen Lectures*,[15] called this composite rule the "Ancient Principle." It received its most precise statement by Robert Smith in *A Compleat System of Opticks* in 1738:

> Any given visible point of an object appears at the intersection of the reflected or the refracted visual ray produced, and of a line drawn through the visible point perpendicular to the reflecting or refracting surface whether plane or spherical.[16]

It is independent not only of the immission and emission theories of vision, as we have seen, but of the distinction made by Kepler, and now in general use, between the virtual and the real image. I should stress that the ancients did not make this distinc-

15. (London, 1669), Lect. 18.
16. Remarks on Article 138, para. 212.

tion. They merely tried to find a rule for locating the apparent place of the object seen in indirect vision. In 1604 Kepler correctly rejected the Ancient Principle, but for such incorrect reasons as that if the mirror at its junction with the cathetus "is covered or blocked off [the image] can nevertheless be clearly seen," [17] in favor of his new rules that still stand. Nevertheless, it survived beyond Kepler. The Jesuit mathematician, Andrée Tacquet (1602–60) described it as "the most fecund in all catoptrics," [18] and, although he witnessed its collapse, based his whole catoptrics upon it. After Tacquet it seems to have just faded away, although Joseph Priestley in 1772 still treated it as received theory.[19]

The Ancient Principle, which was central in classical optics for all of its two thousand years, is properly regarded as a special limiting case of Kepler's, and therefore of modern, principles. The cathetus can be regarded as one ray, and any other ray can be used to find the locus of the image. The Principle, while it applies to reflection or refraction at a single surface whether plane or spherical, cannot be applied to a lens other than a monocentric lens because the normal to one surface will not be the normal to any other surface.[20]

What is the counterpart in classical optics of Kepler's Rule (2) which states that the eye is a sort of camera with a converging lens and screen that takes inverted pictures of external objects? It is little wonder that his momentous discovery of the inverted retinal image so captured the imagination of philosophers that, from his day to ours, the dominant theories devised to explain how we get to know the external world are modeled upon it. The mind imprisoned in its brain receives pictures sent to it from the external world. Most representative or copy theories of perception, including most sense-datum theories, take their origin, so it seems to me,

17. *Supplement to Witelo*, Ch. 3, 1, my translation.

18. *Catoptrics*, Bk. I, prop. 22, from his *Opera Mathematica* . . . (Editio Secunda . . . Antverpiae, 1707), my translation.

19. *The History and Present State of discoveries relating to Visible Light and Colours* (London, 1772), pp. 12 ff. Priestley treated Ibn al-Haitham, A.D. 1000, and Alhazen, twelfth century, as different persons.

20. For valued discussions on the Ancient Principle I am indebted to Professor Rudolf Kingslake of The University of Rochester and Eastman Kodak Company.

from the "discovery" that the eye is a camera, a machine for taking pictures of the external world.

In classical optics there are two solutions to the problem of the mechanism of vision in the eye corresponding to Kepler's solution. Of these, one represents the emission, the other the immission theory. The success of the search for the correct solution required a solution to the mysteries of the anatomy and the dioptrics of the eye, that is, a solution to the problem of how the eye forms an image. The Greeks, already knowledgeable in geometry but stifled by ignorance of anatomy and lenses, set out on the wrong track by following the dominant emission theory. For in the emission theory no images are required. Galen, for example, in the second century A.D., regarded the crystalline lens as the chief instrument of sight, and the retina (*amphiblestroeides*, net-like tunic) as a conveyor of the visual spirit (*pneuma*) from the brain to the eye. As for the advocates of the immission theory, on the other hand, the search might have ended six hundred years before Kepler if Ibn al-Haitham had not taken it for granted that the image in the eye had to be erect. He realized that "the act of vision is accomplished in such a way that the visual image caught by the crystalline lens, is handed over to the optic nerve." [21] He also realized that the eye functioned exactly like a camera obscura. "If lights from candles are set up opposite the opening of a dark chamber . . . the image of each appears . . . upon the back wall . . . and all this is to be understood as being valid for any other transparent body, including the transparent parts of the eye." [22] He admitted that this was not discovered by him and perhaps not even by the Arabs: "*et nos non invenimus ita*" [23]; and although he placed the image at the center of the eye, he made it inverted. But because he could not accept the notion of the receipt by the optic nerve of an inverted image, he suggested double inversion: "The image cannot proceed from the surface of the crystalline lens to the optic nerve along straight lines and still preserve the proper order of its parts. For all the lines intersect at the center of

21. *Opticae Thesaurus*, Bk. I, Ch. 4, theorem 25; translated by S. L. Polyak, *The Retina* (Chicago, 1941), p. 122.
22. *Opticae Thesaurus*, theorem on p. 17 from S. L. Polyak, *The Retina*, p. 133.
23. Ibid.

the eye and if they continued straight on, their position beyond the center would be reversed; what is right become left and *vice versa;* what is up would be down, and down up." [24]

Thus the classical counterpart of Kepler's Rule (2) is a vague and inaccurate account of the same thing. The inverted retinal image is absent, but the chief insight, that the eye is just another optical system, is present. Long before Kepler, however, it was realized, again by an Arab, Ibn Rushd, in the twelfth century, that the retina was the true photoreceptor: "Since colors and shapes are impressed upon it, the retina is therefore the proper instrument of vision." [25] But Kepler was the first to make the complete synthesis and to take the final step of accepting the theory that the image need not be erect for erect vision. This does not mean that he or his followers devised a theory adequate to explain how erect vision is achieved in spite of the inverted image.

Let me summarize this comparison of the modern or Kepler's and the ancient solutions to the problems set by the Greeks. In direct vision we make use of triangles or cones: with two eyes we locate the object at the apex; with one eye the ancient answer consisting of the apex in the eye gave way to Kepler's with the apex in the object. In indirect vision the Ancient Principle, which describes how to locate the image by using the cathetus and the visual ray, was absorbed by Kepler's "New Principle" which utilizes any two intersecting rays. In regard to the functioning of the eye, the obstacle of the inverted retinal image held up the final solution for almost a millennium. This was so even though the ancients ultimately regarded the eye as an optical system. The history of the search for the correct answers which I have sketched might have been shortened. In its later chapters it is a sorry tale of Western prejudice against Arab genius in which obstacles, first removed by the Arabs, were promptly restored by Europeans.

The striking thing held in common by the ancient and modern solutions was the use of the geometrical method considered first, as the interpretation of the lines and angles of geometry as rays

24. *Opticae Thesaurus,* Bk. IV, Ch. 3, theorem 12. For this translation I am indebted to Edward Rosen.

25. *Colliget* . . . (Venetiis, 1553), II.15, my translation. The Latin is: "Quia colores et formae imprimuntur in ipsam [retinam]; ergo ista tela est proprium instrumentum. . . ."

of light reflected or refracted, and secondly, as the deductive method of Euclid. Thus in the thirteenth century Grosseteste said that "the usefulness of considering lines, angles, and figures is the greatest, because it is impossible to understand natural philosophy without these," [26] while his contemporary, Witelo, said that "every mode of vision can be treated by mathematical or natural-scientific demonstration." [27] But in order to complete the explanation, it was necessary to hypothesize about the nature of the physical thing that moved along the straight lines and about its behavior in reflection, refraction, and diffraction. This construction was essentially mechanical. Both the corpuscular and wave theories were suggested by the ancients: the first by the Atomists and their followers among the Arabs including Ibn al-Haitham; the second by Aristotle among the Greeks and by Grosseteste in the Middle Ages. A mechanical model was thus imposed on the geometrical. Descartes, for example, asked his readers to think of light as consisting of a very rapid movement along straight lines exactly as motion is transmitted along the sticks held in a blind man's hands. This motion of light transmitted to the nerves, although quite different from color, he said, was the cause of our seeing color.

But also in both the ancient and modern accounts psychological explanations were unsystematically mixed with geometrical ones. Some of these have been noted. Such cues as visible faintness and clarity, confusion and distinctness, and intervening objects overlap both accounts. But it seems as though these psychological explanations were, in the minds of these theorists, regrettable stopgaps used temporarily as expedients until the proper geometrical solutions could be devised, for the ideal was a complete geometrical and mechanical theory of vision.

There is no doubt that both the ancients and Kepler regarded their treatment of vision in a geometrical way as a theory of vision. The aim was to explain how we see. Slowly, since Kepler, other aims have crept into the science of optics, but although modern texts largely ignore the treatment of direct vision, the rules they provide still purport to describe what will be seen if certain condi-

26. Quoted by Crombie, *Grosseteste*, p. 110.
27. Quoted by Crombie, *Grosseteste*, p. 215.

tions are fulfilled, and they presuppose Kepler's rules for direct vision. In the light of the conclusions arrived at from the earlier account of method in Part One, I shall now examine the main features of the ancient and the modern theories considered as theories of vision.

CHAPTER VII

The Geometrical Model

1. Features of Geometrical Theories of Vision

IN ORDER to bring out the contrast between these geometrical theories of vision and my own alternative theory I shall abstract five general features from the former. The first two, having to do with procedure in treating vision, were shared by the ancients and the Keplerians. The last three, having to do with the process of vision, were probably peculiar to the Keplerians.

A common feature of the two theories was a large portion of their geometrical content, roughly defined by the remarks of Grosseteste and Witelo just quoted, and which may be paraphrased: Optics is to be treated by the geometrical method; and optics is to be treated by using lines and angles, because without these no natural science can be understood. Writers on optics from Aristotle and Euclid to Kepler and Descartes took it for granted that optics fell under its superior science, geometry. The question is how far it fell under or, in other words, to what extent the geometrical model was applied to optics. The first degree of application was the use of the geometrical method. But we have seen that the phrase "the geometrical method" was used ambiguously. First, it was used to mean merely the use of deduction to set up a system, in which case how the arbitrary symbols of the system were interpreted was irrelevant to the purely formal development of it. If this meaning is given, then the geometrical method holds a monopoly on optics, because this method holds a monopoly on science. It is, indeed, a defining feature of it.

Secondly, however, the phrase was used to mean the geometrical way in which the symbols were interpreted, specifically as lines and angles, etc. In which case, the geometrical method holds no monopoly on optics. Because lines and angles may be used to explain the facts, it does not follow that they must be used. Yet this, too, was taken for granted. Galileo's remark that the vast book of nature is written in mathematical language whose letters are triangles, circles, and other geometrical figures, reinforced the dogma. Because Euclid fathered the geometrical method in the sense of demonstration from principles, and because it happened to be the case that he also used geometrical terms in his science of optics, his followers for the next two thousand years mistook this association for identity. But as the father of deductive systems, instead of using terms denoting lines and angles, he might have used terms denoting smells and tastes. The former are devices invented for the sake of treating various facts in geometry and optics expeditiously.

The third feature manifests a yet further application of the geometrical model. I am not sure that it was present in ancient optics. Certainly it was in seventeenth-century optics and implicitly at least in continental philosophy of the same period. It was also present in the optics of Kant if the optical theory implicit in his *Critique of Pure Reason* is abstracted from it. Not only was geometry used as the model for explanation in general in optics, not only were the symbols used interpreted geometrically, but geometry was imposed on the actual manner in which we see. A phrase shared by Descartes and Leibniz, already quoted, characterizes this feature. Both claimed that the process of seeing objects near to us involves a reasoning or reckoning from lines and angles in virtue of our possession of the rudiments of "a natural geometry" (*ex geometria quadam omnibus innata*). The mechanics of this process is shown by the relationship existing between Kepler's Rule (2) and his Rules (4) and (5), or between Axioms VII and VIII of Newton's *Opticks*. Since the eye is an optical system with a converging lens and screen then, given the distance and size of the object and the focal length of the lens, we are able to construct the exact size of the retinal image and, conversely, the distance and size of the object (Rules (1) and (2)). There is no difficulty in this. Kepler was able to do this on paper. Then, however, he made the amazing leap to the conclusion that the eye or, rather, the *sensus visus,* could solve the converse problem. Given the size of the retinal image, we are able to reason backwards to the distance and size of the object by using the triangle with its base in the pupil and apex in the object (Rule (4)). Then, if mirrors or lenses are placed in front of the eye we apply the same technique, ignoring what happens to the rays whether reflected or refracted, and locate the object at the apex of the triangle, that is, at the place from which the rays appear to diverge in falling upon the eye (Rule (5)).

It may be the case that this is what happens when we see. Equipped with this natural geometry we utilize it constantly, although for the most part we are unaware that we do so, just as, as Descartes said, when we squeeze a body with our hands we notice the body, not the movements of our hands. We utilize triangulation just as a surveyor does, although while we ignore the process, he is conscious of it, and while we do it largely inaccurately, he does it accurately. But it seems likely that Kepler and his followers confused two very different things. While examining the methods of Newton and Descartes I argued that they confused properties of the deductive procedure with properties of the physical process. Having *decided* to put necessary connections between principles and consequences in their explanatory procedures, they thought they were able to *find* necessary connections between active principles and their effects in the explained physical process. In a similar fashion, so it seems to me, Kepler superimposed procedure upon process. Able to demonstrate how we see, he concluded that when we see, we demonstrate. Having successfully used two features of the geometrical model to explain the facts of vision, and being quite enthralled by his geometrical zeal, he used a third. Thus a spectator feat became an actor feat; an explanation of how we see was fused with how we see; knowing *that* was fused with know-*how.*

The extended application of models may be of value. In the use of models or metaphors, the main conditions are utility, including illustrative power, and awareness. I repeat that it may be the case that we do see cups and saucers as we compute the distance of stars, or as we deduce a conclusion in logic or mathematics, and that brutes and children, when they see something draw near or recede from them, do this in virtue of natural geometry. The facts of the matter are very hard to find. Accordingly, whether Kepler's is a good account or not depends on consequences not yet told. As for awareness, it may have been the case that Kepler and his followers *were* aware of the extended application of their model. The citizens of Oz may not have been duped after all but may have shared the Wizard's awareness that the greenness of their beautiful city was due to their green spectacles. Descartes, for example, guardedly said that we see distance *"as it were"* by natural geometry. I conclude tentatively, therefore, that while this additional application of geometry to vision looks like pedantry, it may still be valuable, and whether it is or not depends on factors not yet considered.

It is difficult to discover whether the ancients imposed geometry on vision in this extreme fashion. The Greeks invented geometry and the deductive procedure, and they used lines and angles to explain optical phenomena. They used geometrical constructions to show how to find the place and apparent place of the object seen directly and indirectly. They did this on papyrus, but did they think that this procedure was duplicated in the visual process? In indirect vision the answer seems to be negative, because the perpendicular let fall from the object had not yet merged into a ray. It was just an instrument for showing the place of what we see—not an instrument that we use to see. Taking this as our cue, it seems that in direct vision also, the triangles (for two eyes) and the cones (for one eye) they used did not have the same function as Kepler's triangles and cones. If this is so, then a possible reason why they avoided an extreme geometry of vision or "natural geometry" was their ignorance of the dioptrics of the eye.

The fourth feature is the rule that in order to know something by means of something else we need not notice the latter. This feature usually accompanied the previous one although the writers on optics were vague about it. Kepler said that we use his triangles as instruments (*adminicula*) when we see. But do we notice them? He said that they can be taken note of (*considerari*). Again he said that "to see is to feel (*sentire*) the stimulation of the retina."[1] Although he used such words, it is difficult to decide whether he meant that we notice these things just as, for example, we notice muscular sensations. Descartes was clearer, but even he fumbled on the difference between what we actually notice and what we infer we use. His general position is expressed in the rule as I have stated it. Every time we see shapes having size, position, and distance, there are corresponding movements in the brain which nature has appointed as a "means for the soul" enabling it to see "without in any way having to know (*connoitre*) or think of (*penser*) " the means.[2] Half a century later, Molyneux used the rule to explain how we see things erect although the retinal image is inverted. "The mind takes no notice of what happens to the rays in the eye by refraction or decussation but hunts back by means of each pencil of rays."[3] It would not help at this stage to deny the validity of this rule directly by saying such things as that surely if we use anything as a means to know something else the former must be noticed. The rule is properly regarded as a resolution, neither true nor false, that may be used to explain the facts of vision. Whether it is adequate or not depends on other considerations.

The final feature is the important role of the retinal image in vision. Seventeeth-century optical writers never failed to refer to the diminished inverted pictures of external objects formed in the fund of our eyes whenever we see. Their discovery was a momentous event in the history of optics. The proof that they exist in living eyes was that they can be seen in dead eyes in just the same way that a real image can be seen on a cinema screen. "When you have seen this picture on a dead animal's eye . . . you cannot doubt that a quite similar picture is produced in a living man's eye."[4] The mere existence of these pictures was thought a strong argument in favor of their being the immediate objects of vision because they did not seem to answer any other purpose. If this was so, then the process of vision must

1. "*Videre est sentire affectam retiformem*," *Dioptrics*, 61, my translation.
2. *Dioptrics*, VI.
3. *New Dioptrics*, p. 289.
4. Descartes, *Dioptrics*, V.

involve inference because now external objects could not be directly seen. But then, at the same time, it was thought that the pictures could be compared with the external objects, for the optical writers said that the two sorts of objects were exactly alike in shape and color but unlike in distance, size, and situation. Thus it was concluded that we judge the shape and color of objects by means of resemblance, their distance and size by computation from the size of the picture, and their erect position merely by ignoring the inversion of the picture. The judgments of the distance, size, and position of objects thus became the traditional problems. The typical answer to the problem of judging distance was given by Kepler's Rule (4), already discussed, but it may be helpful to give a nineteenth-century version. Helmholtz's was as follows:

> The same object seen at different distances will be depicted on the retina by images of different sizes and will subtend different visual angles. The farther it is away the less its apparent size will be. Thus, just as astronomers can compute the variation of the distances of the sun and moon from the changes in the apparent sizes of these bodies so, knowing the size of an object . . . we can estimate the distance from us by means of the visual angle subtended or, what amounts to the same thing, by means of the size of the image on the retina. . . . By variations of the retinal image . . . one eyed persons are able to form correct apperceptions of the material shapes.[5]

There are problems in this, the accepted theory about retinal images, that optical writers, it seems, have not been sufficiently aware of. One is the problem of their nature. Are they colored pictures that can be seen, or are they geometrical entities— the supposed meeting places of converging rays—or are they inferred physical events on the physical retina? The same question can be asked of any optical image. Since Descartes, Newton, and others adopted Kepler's name "picture" for them, and since they said that they could see them, it seems that they thought

5. *Handbook, 3,* 282, 297; see James P. Southall, ed., *Helmholtz's Treatise on Physiological Optics* (Op. Soc. of Am., 1925), Vol. III, *The Perceptions of Vision.*

that there are colored pictures in our eyes that we use to see with. The problem of the nature of the optical image, to which Professor Ronchi addresses himself,[6] I shall take up in the next chapter.

Descartes, it is true, avoided some of the difficulties I have been discussing, but by doing so, he ran into others. He saw clearly that if the picture in the bottom of the eye was regarded as the immediate object of sight, which by means of its resemblance to an external object makes us aware of the latter, then it would be "as though we had another pair of eyes to see it, inside our brain."[7] He avoided this by pushing the resemblance, as it were, further back. We see by means of the movements in the brain that resemble external objects "structurally" (my word). But at the same time he held that there are colored pictures in our eyes which copy external objects because we can see them in other eyes, and that these pictures are "transmitted to the inside of our head";[8] and he also held that we see by using the different shapes of the single eye and the mutual relation of the two eyes. These instruments or means are the same as the triangles described by Kepler. Thus he retained the pictures in the eye although he said we do not use them, and he retained images in the head which act directly on our souls as the means by which we see although we do not notice them. This pre-ordained harmony instituted by nature between movements in the brain and the external world leads to difficulties which I shall mention shortly.

In summary, there were five general features of seventeenth-century optical theories which I have taken as the commonly received opinions. The first two were also commonly received in ancient optics: (1) the geometrical method considered as the use of deductive systems; (2) the geometrical method considered as the use of geometrical symbols in the principles; (3) the application of geometry to the process of vision, i.e., we see by means of a natural geometry; (4) the rule that the means used in vision need not be noticed; and (5) the rule that retinal images considered as

6. *Optics: The Science of Vision,* Ch. VI.
7. *Dioptrics,* VI.
8. Ibid.

colored pictures are the immediate objects of vision which we use to make inferences to external objects. It will be apparent from my earlier account of the Linguistic Theory of Vision that of these five features I retained only the first in the Linguistic Theory. I was able to do this because I concluded that the use of the geometrical method in this sense is not peculiar to geometry or geometrical optics. The second was replaced by the decision to use symbols that denote items that many regard as items of actual experience such as colors and muscular sensations. The third and fourth were replaced by the decision to use the linguistic model to illustrate vision. Replacing the notion that the process of vision involves inference, including computation from data that are not actually data to the perceiver because they are not necessarily noticed, was the resolution to adopt the view that the way we see things in space is exactly like the way we understand an ordinary language. The fifth was replaced by the decision to adopt the view that colors are the immediate objects of vision.

2. Nativism and Empiricism

The third and fourth features of what I shall call the Geometrical Theory of Vision define what might be called a Nativism, Intuitionism, or Rationalism in optics, while the corresponding features of the Language Theory define what might be called an Empiricism in optics. The distinction between these rival theories cannot be that between the geometrical method (deduction) and induction, the Nativists using the former and the Empiricists the latter. Similarly, the Continental Rationalists, Descartes, Spinoza, and Leibniz, cannot be distinguished from the British Empiricists, Locke, Berkeley, and Hume, by the respective use of the deductive and the inductive methods; for Rationalists hold no monopoly of what was traditionally called the method of synthesis, nor do Empiricists hold a monopoly of the method of analysis. Nor can the distinction be based on the manner in which the arbitrary symbols admitted into the premises of the respective deductive systems are interpreted, the Nativists using fictions like lines and angles, but the Empiricists using observables like colors and feels. Nativists or Rationalists hold no monopoly of the use of mathematical hypotheses in explanation, although it is true that Empiricists prefer to associate themselves with the use of terms denoting observables in their premises. The distinction must be based, therefore, on other factors, and these are now at hand in the third and fourth features I have isolated. Nativists or Rationalists think that we can acquire knowledge of the world by the deductive method; for example, in their optical theories they hold that we see by natural geometry, using a kind of natural computation; we know *how* as we know *that*. Empiricists, on the other hand, hold that we come to know the world by the slow steps of experience; for example, in their optical theories they hold that we learn to see as we learn any other skill involving learning how to respond to a host of similar stimuli; a sharp distinction is made between knowing *that* and knowing *how*, between explanation of how we see and knowing how to see.

There are several kinds of Nativism in optics of which I shall isolate three, calling them respectively, Common Sense, Cartesian, and Kantian. The first was defined by Bailey: We see things in space, as it were, "straight off." "We directly and intuitively see objects. . . . This is a simple perception of which no analysis can be given . . . the fact itself cannot be disputed. . . . Metaphysical investigation and physiological inquiry . . . confirm the universal belief of mankind in the direct visual perception of the three dimensions of space." [9] Another version of the Common Sense view was given by Porterfield. "The judgments we form of objects being placed without the eye, that is, of the situation and distance of visual objects, depend not on custom and experience but on original connate and immutable law to which our minds have been subjected from the time they were at first united to our bodies." [10] The second or Cartesian I have already described. It differs from the first in that vision is not direct, but involves inference in virtue of the supposition that there is a pre-established harmony between brain events and external objects. The third or Kantian was defined by Abbott: "This co-existence [of objects seen]

9. Samuel Bailey, *A Review of Berkeley's Theory of Vision, designed to show the unsoundness of that celebrated speculation* (London, 1842), pp. 105, 237–38.
10. William Porterfield, *A Treatise on the Eye* (Edinburgh, 1759), Vol. II, scholion, p. 299.

demands the idea of space as the indispensable form of its intuition. . . . The perception of distance (as distinct from the measure of it) . . . is the natural concomitant of vision." [11] Thus this variety contains some empirical elements. We see things in space straight off, but experience is required to localize them.

These remarks show the general character of Nativist theories. All of them involve, I think, answering the Molyneux problem[12] in the affirmative. If direct tests, such as those made on people born blind and made to see, are not considered as capable of testing the soundness or unsoundness of Nativist theories, it is difficult to discover what sort of test would be capable at this stage. If this is true of Nativist theories, including the Geometrical Theory described, it is also true of the Linguistic Theory I am advocating. The general features found in all are most appropriately regarded as being neither true nor false but as decisions or resolutions. In which case the claims made by the advocates of the Geometrical Theory that lines and angles are necessary to explain how we see, that we have some innate contrivance enabling us either to be aware of the retinal image or to see things in space as a result of brain movements, and that the means we use to see distance, size, and situation need not be noticed, turn out to be decisions or even pretences of the form: "Let us pretend that these are true in the hope that we can explain or illustrate the facts of vision." In the same way the claims made by the advocates of the Linguistic Theory that vision is a language, whose signs, noticeable by us, we must learn to interpret by the slow steps of experience, are of the same order. There is one factor, however, that can be mentioned now which swings the balance slightly in favor of the Linguistic Theory as a scientific theory. The decision that the means used in vision are not noticeable tends to stifle inquiry at the outset by stopping testing, whereas the contrary decision allows it to breathe.

3. Ronchi's Theory

Before devising tests that may enable us to decide between the Geometrical and the Linguistic Theories, I shall make some gen-

11. Thomas K. Abbott, *Sight and Touch: An Attempt to Disprove the Received (or Berkeleian) Theory of Vision* (London, 1864), pp. 81, 162.
12. See above, Ch. V, sec. 2.

eral remarks about another theory which overlaps the Linguistic Theory in some important respects. These will enable me to clarify further the unique character of the Linguistic Theory. Ronchi is critical of the present state of modern optics.[13] Inappropriate naming breeds confusion of thought. He undertakes to redefine "optics" more in accord with its etymological meaning, primarily because he wants to define the boundaries of a new science. To talk of the "optics of X-rays" or the "optics of electromagnetic waves" is to abuse the word "optics." There are three means for detecting radiation: photoelectric cells, photosensitive emulsions, and eyes, whose reactions are indicated by an electric current, a blackening, and perception of light and color, respectively. These means should issue in three separate but sister sciences: photoelectricity, photography, and optics. Since the first has nothing "optical" about it, and the second nothing "optical" in so far as the taking of the photograph is concerned, both should be banished from the domain of optics. Physical optics is nothing but a part of radiation physics. Since geometrical and wave optics are "now deprived of a secure physical basis and reduced to the status of provisional studies of schematic models" by the conscious admission that the correspondence with experience does not exist, they lose their standing as branches of optics and become "chapters of mathematics." Finally, since physiological optics, in practice, is mainly concerned with the functioning of the eye, the word "optics" is also abused here.

According to Ronchi most of us are thoroughly confused over the meaning of the term "light." We use the same word to refer to the light that is seen and the invisible physical light supposed to continue to exist in the absence of eyes. He would therefore revive the ancient distinction between visible *lux* and invisible *lumen*, leaving the latter for treatment in physics. Similar remarks apply to color. To avoid confusion, he would have us preserve the distinction insisted upon by Newton when he said that the rays of light are not colored: they are color-making.

From considerations like these Ronchi concludes that modern optics has lost its reason for existence as an autonomous science.

13. For details sketched in this section see Vasco Ronchi, *Optics: The Science of Vision*, Ch. I.

His book, *Optics: The Science of Vision,* as its title suggests, defines the limits of a new science of optics whose main purpose is the same as that of ancient optics: to explain how we see. It is an inter-disciplinary science which uses the contributions of physics, physiology, and psychology, for in every optical situation there are three corresponding phases: the physical, the physiological, and the psychological.

It might seem, then, that the only difference between Ronchi's new optics and recent physiological optics is one of naming. But the difference in names indicates a difference of emphasis and direction. As Ronchi develops his subject, the observer with his memories, prenotions, etc., emerges as "the absolutely predominant factor" or principle used to explain how we see the external world. Since he believes that the psychological phase dominates the other two in every optical situation, psychological laws are therefore the foundations of the science of vision. Because he not only reinstates the role of the observer in optics but treats it as the dominant one, and because the word "optics" has meaning for him only when the aim is to discover the conditions and laws that permit an observer to see and see well, he calls his discipline "anthropomorphic."

If these features distinguish the optics Ronchi so cogently argues for from physiological optics, their common features, supposing that the latter neglects psychological elements, distinguish both disciplines from what I am advocating. I am advocating a theory abstracted from the physical and the physiological. In other words, although the whole optical situation may contain three phases, it is possible to ignore the first two and base a theory on what, most people would probably agree, the observer either actually experiences or can be taught to experience. He does not experience radiations of energy, matter conceived as an assemblage of corpuscular units, light conceived as *lumen,* rays or waves converging and diverging, the events on his retina, and so on. But he does experience colors of various hues, blurred or distinct, faint or clear, and he can experience the turn of his eyes as he watches an object approach his nose and the strain that accompanies trying to see distinctly an object a few inches away. He can see intervening objects, and he can feel his eyes turn up and down, to the left and right. He can remember how things looked and felt in the past, how long it took to reach them or run away from them, and so on. And he can imagine what they would feel like if he were to try to feel them, how many "distances of time" they are remote from him, and so on.

The items just mentioned almost exhaust all the cues we need in order to tell, by looking, the distance, size, and situation of objects. We use these cues to enable us to see things in space in just the same way as we use cues to enable us to understand words. Accordingly, the terms denoting them appear in the principles of the Linguistic Theory of Vision.

CHAPTER VIII

Test Cases

1. The Barrovian Case

WE CAN DEVISE TESTS that will help us to judge the merits of the Geometrical and the Linguistic Theories. Of these, I shall isolate three celebrated cases of visual illusion, the solutions of which remain controversial: the Barrovian Case for distance, the horizontal moon for size, and the inverted retinal image for situation; and I shall treat them in that order. Let us see first, however, where the argument now stands. We know the main rules of both theories, and we have decided that both sets are designed to solve the same problems—problems that were set by the Greeks. I concluded that it was difficult to choose between the two theories by a direct examination of the rules. Accordingly, in the following tests we pretend that we do see by geometry or that we do see by language as the case may be. But if one theory cannot accommodate what are generally agreed to be the facts while the other can, or if one shuts the door on the invention of further tests while the other opens it, it may be reasonable to keep up the pretense that we see either by geometry or by language but not by both.

The Barrovian Case is called after Isaac Barrow, the teacher of Newton who, in the last of his eighteen lectures on optics,[1] encountered "a certain odd and particular case." In it, he said, "something peculiar lies hid, which being involved in the subtlety of nature will, perhaps, hardly be discovered till such time as the manner of vision is more perfectly made known. Concerning which, I must own, I have hitherto been able to find out nothing

---

1. *Eighteen Lectures.* Unless stated otherwise, all quotations are from Lect. 18, translated by George Berkeley, *An Essay towards a New Theory of Vision,* 29.

that has the least show of probability, not to mention certainty."
Unable to solve it, he concluded his lectures by issuing a challenge which he left as a rich legacy to his students: "I shall,
therefore, leave this knot to be untied by you, wishing you may
have better success in it than I have had."

The case is concerned with indirect vision and specifically the
apparent place of the object seen through converging lenses and
mirrors, and thus with the application of what I have called
Kepler's Rule (5): the optical image seen by reflection or refraction is located at the place from which the rays appear to diverge.
Textbooks of geometrical optics from Kepler's day to ours show
various graphical constructions of the images formed by converging lenses and mirrors, virtual when the object is placed inside the
focal point, and real when the object is placed outside the focal
point. To quote from a recent text, "A virtual image cannot be
formed on a screen. The rays from a given point on the object do
not actually come together at the corresponding point in the
image; instead they must be projected backward to find this point.
. . . [In Fig. 4B] the point Q' is said to be a virtual image of
Q since when the eye receives the refracted rays, they *appear to
come* from a source at Q', but do not actually pass through Q' as
would be the case if it were a real image." [2] In the construction of
virtual images it is customary to sketch in the eye of an observer,
while in the construction of real images it is customary to omit
it. This suggests that while the existence of virtual images is
thought to depend on the presence of observers, that of real images
is not. It seems at first that Kepler's rule, illustrated by these constructions, fits experience. It seems that we confirm it every time
we shave or powder our faces with the aid of a plane mirror, and
every time we notice the phenomenon of the bent stick seen
through water.

But the modern texts ignore one case, that in which the object
is placed beyond the focal point of a concave mirror or converging
lens while the eye of an observer is placed between the system and
the real image (Figs. 8A and B). What, according to the laws of
optics, are we supposed to see? Or, as Barrow, who was considering

2. F. A. Jenkins and H. E. White, *Fundamentals of Optics* (2nd ed. New York,
1950), pp. 20, 42, my italics.

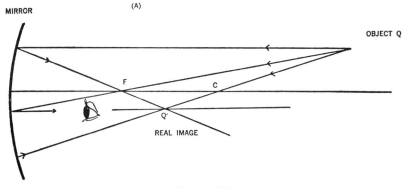

Figure 8A

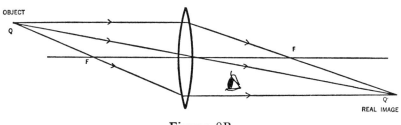

Figure 8B

the object as a point, asked: "The question now is, where the point Q ought to appear." Barrow, following Kepler's rule, immediately answered his own question: "If we exclude all anticipations (*praenotiones*) and prejudices (*praeiudicia*)," the object ought to appear "extremely remote" because diverging rays mean near, less diverging rays mean less near, parallel rays mean far, therefore, converging rays mean very far. Barrow's reasoning so far was impeccable. Compare it with Professor Ronchi's: "If the famous fundamental hypothesis of seventeenth-century optics were confirmed, the observer should see the image nearby when the waves are divergent; he should see it at infinity when the waves are plane; and when the waves are convergent, who knows what he is supposed to see?" [3]

This was what Barrow argued ought to happen according to

3. *Optics: The Science of Vision*, sec. 195.

the rules. But what did he find happening when he looked himself? He found that the facts were directly opposite to what they ought to have been. The point, he said, does not appear behind the head in the place of the real image because that would be "contrary to nature." He noticed that if he placed his eye right up against the lens or mirror, the object appeared nearly in its own natural place. Then, as he drew his eye slowly back, he noticed that the object seemed to draw nearer and nearer until, at length, when the eye approached the place of the real image, "the object appearing extremely near begins to vanish into mere confusion" (cf. *PC,* 501).

Barrow's observation report was good. Although he was completely nonplussed by this turn of events, he remained undismayed. Having found the facts "repugnant" to the theory which, he said, "I know to be manifestly agreeable to reason," he refused to renounce the theory—the theory of modern as distinguished from ancient optics.

Is the case paralleled in ancient optics? Did anyone notice a similar discrepancy between theory and fact? As I have noted, the ancient rule corresponding to Kepler's for locating the apparent place of the object seen through mirrors and lenses was what Barrow called the "Ancient Principle." Apart from Bacon, who noticed that this principle did not seem to fit some cases, the only theorist, to my knowledge, who isolated the parallel to the Barrovian Case was Tacquet, who wrote after Kepler. Having used the Ancient Principle, "the most fecund in all catoptrics," to explain the things seen in plane and convex mirrors (to repeat), "any point of the object appears nowhere else than at the intersection of the reflected ray with the perpendicular of incidence," [4] he then encountered an instance of the Barrovian Case. For spherical concave mirrors, he offered the following theorem:

> If the eye is between the center and the mirror, then objects placed below the center make two images: an inverted one between the center and the mirror, and an erect one beyond it.[5]

4. *Catoptrics,* Bk. I, prop. 22, my translation. Future references to Tacquet are from the *Catoptrics.*

5. Bk. III, prop. 29, my translation.

In the diagram (Fɪɢ. 9), which copies Tacquet's, it is seen that the first image, "real" in modern optics, is apparently accounted for by the Ancient Principle, "but the second, though established by a sure experiment (*experientia certa constet*) cannot be demonstrated from it, because the image repeatedly appears outside the

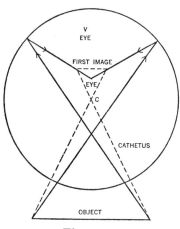

Figure 9

junction of the reflected ray with the cathetus." [6] In the face of these facts, Tacquet did not renounce the Ancient Principle. Instead, he said: "In concave mirrors we postulate this only so far as its truth reveals itself." [7] In this state he left the matter.

Tacquet's rule, being the same as Euclid's, is a special case of Kepler's. If the latter cannot fit in the facts, neither can the former. The second image that Tacquet and Barrow saw is, indeed, inexplicable on the principles of geometrical optics, ancient or modern, for, in terms of the one it is outside the junction of the cathetus and the reflected ray, and, in terms of the other it is neither *real* (the rays do not actually pass through it) nor *virtual* (the rays cannot be projected to pass through it). According to

6. Bk. III, props. 29 and 30, my translation.
7. Bk. II, prop. 22, my translation.

the Geometrical Theory of Vision, therefore, this image or effigy does not exist. Yet Barrow and Tacquet saw it, and anyone can confirm its existence by carrying out a similar experiment. Most of us have done so on every occasion that we have looked through a magnifying glass. The Barrovian Case thus subverts Kepler's Rule (5), and by the same token, Newton's Axiom VIII, considered as a principle designed to show what we see. But Kepler's rule is the application to indirect vision of his rule for direct vision, his distance-measuring triangle. Here then is a case in which Kepler's triangle cannot apply because there is no vertex for it. Yet, although it is blurred, we manage to see the object. Accommodation, interpreted as an innate mechanism, cannot be the telemetric device we use because the eye cannot accommodate converging rays. Even if, for the sake of carrying out tests, we pretend that we see by geometry, this case is beyond its reach: the Barrovian Case is a disconfirming instance of the Geometrical Theory considered as a theory designed to explain how we see.

What determines what we see? Optical theorists, three hundred years after Barrow, ignore the problem. Can the Linguistic Theory fit in this phenomenon while the Geometrical Theory cannot? Let us see how we may begin to apply the language model. We know from the way we understand the words of an ordinary language that by constant experience an habitual or customary connection is established that enables us to bridge the gap between words and what they mean. We know that the signs we use are noticeable although usually overlooked. In the odd and particular Barrovian Case we have, then, to find the overlooked signs. From our understanding of language we can infer that if an American were to meet a foreigner who used English words in a directly contrary signification, the American would not fail to make contrary judgments of the foreigner's meanings. For example, if he were to meet a Russian diplomat of the thirties who used "yes" to mean no, "democracy" to mean tyranny, "peace" to mean war, "near" to mean far, and "far" to mean near, without giving any indication of his different uses of these words, it might take the American twenty years to see through this double-talk to the real meanings (E, 32).

Now the Barrovian Case is exactly like this double-talk. It involves not a new language with new terms but an old language with new meanings that have to be re-learned. The observer is placed, not in the predicament of a foreigner hearing a strange tongue for the first time, like the man born blind and made to see, but in the predicament of a citizen hearing his own tongue used in a strange way. All we have to do then is to discover those signs previously established by custom to signify one thing, now being used to signify another. The facts of the case are pretty much as Barrow described them. The object *seen* is fuzzy or blurred, and as the object or eye is moved farther and farther away, the fuzziness increases while the object appears to draw nearer and nearer. I have decided, it is part of my theory, that we do not see distance directly. There must be, therefore, a sign that suggests it to us. We have already isolated visible fuzziness or confusion as an item constantly associated with near distance, so that if no other one is offered, then, when he *sees* this fuzziness, the observer cannot help but see a near object. In other words, the object "speaks" with well known "words"; they are fuzzy or blurred, but this time they suggest a directly contrary meaning. It follows that the observer will take these "words" in the same sense as he has always done and will be unavoidably mistaken (*E*, 31, 32).

By correlating these visible confusions with different states of the retinal image, we can illustrate both the adequacy of the solution just offered and the corresponding difficulty of the Geometrical Theory. I can do this while remaining neutral on the question of whether the retinal image is a colored picture, a geometrical entity, or an inferred invisible physical event. The three diagrams[8] (FIG. 10) show a normal eye receiving (A) parallel rays accurately focussed on the retina; (B) rays so diverging that they focus beyond the retina; and (C) rays made to converge so that they focus before the retina. According to the Geometrical Theory, as Barrow argued, if we exclude "all prenotions and prejudices," i.e., all other factors, then, if (B) occurs, the observer infers near; if (A) occurs he infers far; therefore, if (C) occurs, he ought to infer extreme remoteness. But Barrow failed to consider an important supposed fact, illustrated in the diagrams (B) and (C):

8. These copy Berkeley's *Essay*, 35, who got them from Molyneux, *New Dioptrics*, p. 103.

the confusion circles on the retina cover the same area in both. According to my theory, the observer cannot notice what occurs on his retina; all he notices is the confusion itself, and this is the same whether the rays are diverging (rays fall converging on the

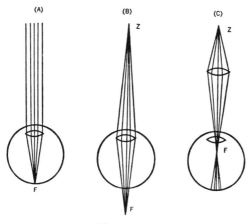

Figure 10

retina) or converging (rays fall diverging on the retina). It follows that, all other cues being excluded, as far as the observer is concerned, the situations illustrated by diagrams (B) and (C) are the same. The object Z is seen at the same distance in the situation illustrated by (B) as in the situation illustrated by (C), i.e. in the Barrovian Case (E, 35, 36).

There are, however, several difficulties connected with my solution, of which I shall briefly treat two. First, as Rudolf Kingslake has pointed out to me in a discussion of the situations (B) and (C): "In either case, the intercept of the ray cone at the retina is a circle of confusion, and the observer cannot usually distinguish between the two situations unless he makes an experimental change in accommodation. This sharpens the image in the case of diverging entering rays and increases the unsharpness for converging entering rays." [9] M. H. Pirenne in a discussion of the two situations concludes that "while the degree of confusion perceived is the same for converging and diverging rays, the sensations may

9. Personal communication; see Ch. VI, note 20, above.

yet differ in some other respect, as in fact they do on account of the chromatic aberration," [10] for example, the color sensations are different. But my solution accommodates the facts here adduced. These facts constitute what may be additional cues that the observer may use to distinguish between the case of nearness directly or ordinarily seen and the case of nearness artificially or indirectly seen as in the Barrovian one. But this is a new test different from the Barrovian. In the latter, as Kingslake says, after accommodation, the fuzziness increases. But accommodation is an unconscious reflex not itself noticed; only the effect is, namely, increasing fuzziness which unavoidably suggests increasing nearness because it has been frequently associated with it in the past. But the two situations are not being compared in the present. Nevertheless, it is probably true that a skilled performer in the second test case, that of distinguishing between the two situations, would not so easily be misled in the Barrovian Case. But then he would be bringing to bear "prenotions and prejudices" which Barrow excluded.

Secondly, if fuzziness is the cue we use, then, "in the Barrovian Case, purblind would judge aright." That is to say, a strong myope would find the object receding from him, because for him, confusion is associated with distance, and distinctness with nearness. But though I am partially purblind with a dioptrical error of −4, I notice that in a rough approximation to the Barrovian Case events unfold much after the same manner as Barrow described. Can my solution accommodate this apparent disconfirming instance of it? I think it can, for the following reasons. When I perform the test I notice that the size of the visual object increases with the increasing fuzziness and nearness. This suggests that visual size, not confusion, may be the cue used to see distance, as Robert Smith claimed it was, not only for the Barrovian but for all cases: "I conclude then by induction of particulars in all these experiments that, whatever be the variations of the divergency and convergency of rays, or of distinctness and confusedness, or of brightness and faintness, the apparent distance of a given

10. "Physiological Mechanisms in the perception of Distance . . . ." *The British Journal for the Philosophy of Science*, Vol. IV, No. 13 (May 1953), p. 17.

system of known objects seen with glasses (and with the naked eye) is suggested to us either principally or solely by its apparent magnitude." [11] This claim collapses when we remember that the case proposed by Barrow was in regard to only one visible point. When several points are considered, or the image is supposed to be an extended surface, its increasing fuzziness will, in that case, concur with its increasing size to diminish its distance, which, roughly, may be inversely as both. But if in the experiment we start off with one visible point, nevertheless, as we proceed, it becomes many. This factor, however, can be excluded by so arranging the test that visual size is neutralized (*V*, 68).

Moreover, it is as if the strong myope constantly carries around with him the distinguishing feature of the Barrovian Case, specifically, visual fuzziness. This condition is alleviated for part of his waking life by suitable spectacles, but whenever he removes them or glances beyond the rims, he gets visual confusion once more. Thus for him both visual fuzziness and distinctness are associated with both nearness and distance. What does he do? He learns to disregard these cues and to rely on more faithful ones, just as the American, previously mentioned, has now learned to see through the double-talk of the Russian diplomat and to rely on more massive cues, such as disclosure by unstudied talk, the whole context of the sentence, and a host of others. But it seems certain that if the observer had never worn remedial spectacles then, all other factors having been neutralized, visual sharpness or distinctness would suggest very near distance to him.

Even if my induction from particulars is faulty in the Barrovian Case, even if the sign is not confusion, this does not disconfirm the Linguistic Theory. It merely means that, having picked the wrong cue, further testing may isolate the right one. Nevertheless, it is more in accord with the Linguistic Theory to say that in most situations, including approximations to the Barrovian Case, we use all the cues we can muster. Indeed, those "prenotions and prejudices" that Barrow excluded furnish the essential cues of the Linguistic Theory which could appropriately be renamed the Expectation Theory of vision.

11. *A Compleat System of Opticks*. Remarks upon article 138, para. 232.

2. *Ronchi's Criticisms*

This is a central point in Ronchi's "anthropomorphic" optics which, though not the same as the Linguistic Theory, overlaps with it in this regard. Although he examines numerous cases that reveal the inadequacy of the Geometrical Theory of vision, he focusses on the Barrovian Case.[12] He first refers to experiments that disconfirm the hypothesis of seventeenth-century optics in its "innermost citadel," the plane mirror, which seems on the face of it to reflect a perfect correspondence between Kepler's rules and the facts. He finds that the rules that appear to hold for close distance fail to do so when the mirror is remote (FIG. 4A). In treating the concave mirror, he singles out for attention the extreme counter-instance of the Barrovian Case. "If the observer is supposed to see the image, he would have to see behind his own head, and that has never happened. On the other hand optics forbids any figure to be seen in front, because such a figure would have to consist of the centers of divergent waves reaching the eyes; and since there are no such waves, there should be no figure" (FIG. 8A). When he considers converging lenses, he again chooses the Barrovian as the most revealing test case. "Let us move the object away from the lens. The waves emerging from the lens must be convergent, with their center behind the eyes of the observer. What is he supposed to see? Nobody says. Yet no figures should be seen on the same side of the lens as the object, because on that side there never are any centers of waves emerging from the lens. Countless users of converging lenses in their eye-glasses have shown for nearly seven centuries that things are exactly the opposite of what is required by seventeenth-century optics" (FIG. 8B).

Having concluded that seventeenth-century optics, which is an instance of what I have called the Geometrical Theory of Vision, has shown itself "utterly inadequate to explain" the facts noticed by anyone who looks at or through mirrors and lenses, Ronchi frames his own hypotheses to account for the same facts. As I have noted, he gives an account of vision using physical, physiological, and psychological factors. Illuminatingly, he blurs the distinction

12. Quotations in this section are from Vasco Ronchi, *Optics: The Science of Vision,* in roughly the following order: secs. 192–93, 89, 71, 91, 121–22, 99, 115, 139.

between dreaming and seeing, or, rather, depicts the latter in terms of the former with something added. In a dream the ego beholds bright-colored figures before it although it receives no radiant or mechanical stimulus from outside. Just as these figures are "created by the mind," so, in his view, vision itself "is like a dream built on the basis of information received by means of external stimuli and the peripheral organ of sight." He calls the object of vision the "effigy" or "model," a nonmaterial thing constructed out of certain data on the analogy of a clay model constructed by a sculptor. In which case the mind, as it were, "depicts" the effigy by "giving it" its luminosity (called in my account clarity) and color.

He regards the problems of placing this effigy at a definite distance from us as "the most difficult problem that the mind is called upon to solve." To achieve this, the mind uses all the factors at its disposal: nerve impulses reaching the brain through the optic nerve; accommodation and convergence; previous knowledge, the memory acting as a storehouse; and, lastly, imagination and the mind's own initiative. Accommodation he finds to be almost completely ineffective as a telemetric mechanism. Useless beyond 4 meters, he even hints that we locate the object first by other means, and then accommodate for it. Convergence, though somewhat more effective, is useless for ordinary observers beyond 25 meters. He concludes that in locating the effigy *seen* not only in approximations to the Barrovian Case but in all other situations, the observer "pays no attention to the information based on accommodation or convergence," but places it at the distance "deemed most reasonable by the mind." In other words the prenotions and prejudices that Barrow tried to exclude are given the overwhelmingly important role in Ronchi's theory as they are in mine. Since prenotions and prejudices vary among observers, so do the constructed effigies. One hangs the object on the lens or mirror, another leaves it dangling in space.

This brief sketch shows that many ingredients of Ronchi's theory not only are translatable into those of the Linguistic Theory but also illuminate them. There are, however, some important differences. Ronchi has resolved to treat colors of different hues, brightness, confusion and distinctness, faintness and clarity, visual size, form and position, all as creations or constructions of the mind. He cannot emphasize this point too strongly. Such a resolution is foreign to my theory. These items are given to us, not by us, although it is true that we give them names. If we create them we might just as well say that while listening to a lecture we coin the words we hear. Accordingly, while the dominant picture of the process of vision painted by Ronchi's theory is of a sculptor *creating* clay models, the corresponding picture of the Linguistic Theory is of a child *understanding* ordinary words.

Ronchi, like many of the seventeenth-century optical theorists and their modern followers, has resolved to mix physical, physiological, geometrical, and psychological factors in his explanation of vision. This is a commendable decision and a necessary one for "a compleat system of opticks," such as Robert Smith envisaged. But, as I am trying to show, this resolution is not necessary to a theory of vision simplified by the adoption of the linguistic model to account for the facts in terms that might once have been called "psychological" before recent psychology borrowed the word for a new use. For the Linguistic Theory was devisable by the sons of Robinson Crusoe who, reared after his death, understood language but remained ignorant of theoretical physics, physiology, and geometry.

When he describes the actual process of vision, Ronchi regards all these factors as data. These are all "information" that the mind "arbitrarily integrates" in its creation of the finished effigy. Nerve impulses reaching the brain are put in the same category with actually remembered associations and actually noticed or noticeable cues. This again is a resolution foreign to my theory. From the latter standpoint it must be a case of mixing the procedure used to explain with the process explained, of mixing knowing *that* with knowing *how*, of fusing the description of how we see with knowing how to see. If we say, when we see a cat at a certain distance, that we do this in virtue of integrating information received from nerve impulses and from the memory of past experiences, then we might just as well say, when we understand the word GATTO, that we use our brains as well as our memories.

These different decisions result in different theories of vision. Doubtless, the two theories are rivals in some areas, and tests

could be devised that might lead to choice between them. But
I am concerned to treat them as allies in other areas. In which
case, I regard Ronchi's account as a vindication of the Linguistic
Theory, especially since he shows conclusively the inadequacy of
the Geometrical Theory to explain the Barrovian Case as well as
the next test case, and since in his own hypotheses devised to explain how we see, he places all the stress on those *factors* that are
the dominant *signs* or *cues* of the Linguistic Theory.

3. The Horizontal Moon

Having considered a test case for distance, I shall now consider
a test case designed to enable us to choose between the Geometrical and the Linguistic Theories on the matter of how we see size.
This is the problem of that celebrated phenomenon, the apparent
size of the horizontal moon. Why does the moon look bigger
near the horizon than at the meridian although the angle under
which the diameter of the moon is seen is not greater in the former
case than in the latter? The solutions to this problem, provided to
their own satisfactions by Ptolemy, Ibn al-Haitham, Witelo,
Bacon, Kepler, Hobbes, Descartes, Malebranche, Gregory, Smith,
Wallis, Huygens, Helmholtz and others are questionable.

One might say that this case hardly tests the merits of the two
theories because the Geometrical Theory was not devised to apply
to remote distances. If accommodation and convergence, considered as distance-measuring devices, are almost useless to a bats-
man facing a bowler less than 21 yards away, they are probably
entirely useless to anyone estimating the distance of the moon by
sight. The same conclusion applies to the estimate of size because,
according to the theory, we use the size of the angle under which
the thing is seen, or the size of the retinal image, plus the distance
to compute the size of objects. As Descartes said, "their size is
judged according to our knowledge or opinion as to their distance,
in conjunction with the size of the images that they impress on
the back of the eye." [13] But as I have noted, Kepler, though not
Descartes, was aware that his triangles were useful only close up.
Therefore, it might seem that if his rules represent the Geometri-

13. *Dioptrics*, VI.

cal Theory, then to test the latter by the case of the horizontal moon is an *ignoratio elenchi.*

The correct answer to this objection, however, is that the moon test still applies. Without finishing the performance of this test I know right at the start that the Geometrical Theory is out of its depth.

In my search for instances that may disconfirm either rival theory, I know at once that the moon test disconfirms the Geometrical Theory because the paradigm solution of this puzzle by the Geometrical Theory must be in terms of the size of the angle under which the moon is seen. I also know that other principles are needed to account for the facts. Now nearly all the optical theorists mentioned did rely on other principles to explain the different apparent size of the horizontal moon. All these other principles can be absorbed by the Linguistic Theory because they rely on factors that are noticeable, and which, therefore, may be treated as signs. Since the Geometrical Theory is admittedly disconfirmed in this instance, and since I know *a priori* that the Linguistic Theory can accommodate most of the solutions so far devised, I could really stop here. But in order to show how vividly the linguistic model illuminates this phenomenon, I shall briefly describe four solutions, three of which are traditional while the fourth is peculiarly linguistic and, at the same time, most economical.

I know as before from the linguistic model that there will probably be much more information at our disposal than at first meets the eye. Though it seems that all we are given is a flat luminous disk that sometimes looks as big as a silver dollar and at others no bigger than a sixpenny piece, there are, in fact, a host of other factors accompanying the phenomenon, potential cues usually overlooked in our snap estimation. Some of the features present seem to be as follows: (1) the visual size; (2) intermediate objects such as woods and fields, houses, and clouds; (3) the flat dome-shaped curvature of the sky; (4) the horizontal situation of the moon in the sky; (5) the faintness of the visible appearance as opposed to its vigor or clarity.

Using Mill's methods of agreement, difference, and concomitant variations, which we all use so frequently in our daily lives, it should be possible to eliminate some of these items. The first to go is (1) visual size, because its neutralization is built into the problem, if we regard visual size as correlated with the size of the angle subtended by the disk measurable by a goniometer, and with the size of the retinal image. The size of the angle is the same for the horizontal and the meridional moons. Yet, to some of us, the one moon looks three or four times larger than the other.

The next to go is (2) the factor of intermediate objects. These objects, associated with greater distance, present in the case of the horizontal moon and absent in the case of the meridional moon, suggest the greater distance of the former, and this, in turn, magnifies the size. This is the first traditional solution. Although the factor involved here may contribute to the solution, it seems to be only accidental or, at least, not the main one. For I can deduce from this solution that if one looks at the horizontal moon from behind a wall, it should appear no bigger than ordinary. But it does. We can deduce that if a large number of objects intervene, then the moon will always look much bigger than usual. But sometimes it does not, as anyone can confirm. Adopted by Ptolemy, the Arab astronomers, Wallis, and many others, this solution that the moon looks bigger because it appears remote prompted Helmholtz to ask what he called the "real question": "Why does the sky look nearer at the zenith than at the horizon?" [14]

This brings me to (3) the second traditional solution. The theory that we imagine the sky as flattened rather than truly spherical was adopted by Robert Smith who said that we see or imagine the distance to the horizon as three or four times greater than the distance overhead, and that, consequently, the horizontal moon looks three or four times bigger.[15] Ronchi adopts this solution.[16] It is subtler than the previous one because it need only claim that the flattened dome of the sky fills our imagination, not our actual vision. In which case it meets the objection that the horizontal moon looks just as big from behind a wall. We can deduce from this solution, however, that since we always imagine the sky flattened, the horizontal moon always appears bigger than ordinary. But sometimes, as I have noted, it does not (E, 76, 77).

14. *Handbook, 3,* 290.
15. *A Compleat System of Opticks,* Article 163.
16. *Optics: The Science of Vision,* sec. 120.

It is hard to separate (4), the horizontal situation of the moon, from the last two solutions. It seems certain that the situations of objects influence our decisions about their size. As we raise our eyes from looking at our feet to the horizontal, objects usually look farther away and larger, but as we raise them above the horizontal, their distance and size diminish. This factor probably contributes to the suggestion of the size of the horizontal moon. But that this factor alone is not enough is likely from the experiment referred to by Helmholtz for use in a different context. When a plane mirror is used to bring down the bright meridional moon so that it is seen as the horizontal moon, it looks "not decidedly larger than the meridional moon." [17] This experiment also counts against all the traditional solutions.

But according to H. Leibowitz and T. Hartman it should look larger.[18] For they claim that the larger apparent size of the horizontal moon is explained by its horizontal situation. Experiments performed with terrestrial objects viewed vertically and horizontally by nineteen adults and nineteen children confirm their solution "that human beings have more experience with objects in the horizontal than in the vertical plane, and thus make a larger correction for an object subtending the same visual angle—for example, the moon—when viewed horizontally." But when I perform Helmholtz's mirror experiment I obtain his results which anyone can confirm. Moreover, assuming that we have more experience of viewing the horizontal than the vertical moon, it follows from their theory that the horizontal moon always looks larger than the vertical one to the same adult. But in fact it varies for me from time to time. Finally, from their claim that "since children have less experience with distantly viewed objects, especially when viewed directly overhead, the magnitude of the moon illusion is greater the younger the observer," it follows that adults should find the apparent sizes of the two moons equalizing out the older they get. But this is discounted by my own experience and by reports of very old people.

17. *Handbook, 3,* 291.
18. "Magnitude of the Moon Illusion as a Function of the Age of the Observer," *Science, 130* (September 1959), 569–70.

Something else is needed, therefore, to account for the phenomenon. We know that it must be subject to variation because the phenomenon itself is. This condition, it seems, is met by (5), the faintness of the visible appearance, which involves a solution that meets the objections raised against the others, while the other solutions seem unable to fit in the stubborn fact that the size varies from time to time. The often overlooked cues of faintness and clarity are rendered easily noticeable on either unusually clear or unusually hazy days. Mountains look nearer and smaller or just the reverse. Faintness and clarity, unlike confusion and distinctness, depend on the amount of light seen, or, if one likes, on the amount of physical *lumen* reaching the eye. Confirmation that faintness is the main factor is provided by the fact that in misty weather the size of the horizontal moon is increased. The fact that it may look big even on a clear evening in one place means only the presence of fog or haziness elsewhere between the eye and the moon. This haziness is not noticeable but the result is. The fact that a gauze placed before the meridional moon to produce faintness does not make it appreciably larger, means only that a cue torn from its context tends to be discounted (*E*, 71, 72).

But in this case, as in all others, it is certain that we use all the cues we can muster, and of these, the most important are undoubtedly our prenotions and prejudices. These include the remembered contexts. The moon is no exception. If I have grown accustomed to her horizontal face, the elimination of some factors will not alter her appearance unless I decide to attend very carefully. Constant attention to the phenomenon of the size of the moon on the part of an investigator changes the facts. To be constantly aware of all the cues is to diminish their influence upon us just as to attend to our feelings is to lessen them. Miss Lucia Ronchi of the National Institute of Optics in Florence, who has given considerable attention over a long period of time to the problem before us, has told me that, for her, the horizontal moon now looks no bigger than the meridional one. The same consequence would probably ensue for most of us for whom the body of a man *does* look just as big whether seen at ten feet or at five, above or below us, through a haze or in clear weather.

That this phenomenon of the horizontal moon strongly confirms the illustrative power of the linguistic model is apparent from the following factors: Here is a case where the signs used suggest quite different things; where we tend to overlook the signs; where the visual size remains the same though its meaning varies; where the visual size is quite inadequate to suggest the size we think of; where a sign used in certain circumstances or contexts has a different meaning when used in different circumstances or contexts; where a sign torn from its usual context is almost meaningless; where a sign suggests ambiguously—faintness suggesting both greater distance and greater size at the same time; and where we do not need to be directly acquainted with the thing signified in order to interpret the sign. The analogues of all these factors are well known to us in the linguistic model. The last is interesting in that it shows how successfully the Linguistic Theory meets an objection made against it at its foundations, an objection that, before I draw upon the reserves of my model, seems unanswerable. The objection is this. If vision depends upon frequently experienced correlations of sight and touch how then do we manage to see things that we have never touched? We do see these things just as clearly as those relatively few things that we touch while we *see*. Therefore my theory collapses. But let me call up the reserves of the linguistic model. We all know very well that we can interpret the words of an ordinary language even though most of them have never been ostensively defined for us. We understand the words "Babylon" and "Bonaparte" perfectly well although their actual designations lie forever beyond our reach. We can see the moon and the Matterhorn quite clearly although we have never been to these places.

Indeed the features in this classic puzzle known as "the moon-illusion" that peculiarly confirm the appropriateness of the linguistic model when used to illustrate vision are those that require for their solution just such secondary and unusual properties of the model that I have found in it. By forcing me to use these properties they enable me to display the resilience or adaptability of my theory. Thus I can deduce that the size of the words of visual language is not enough to suggest the size we see or think of, that the

moon and sixpence, for example, may have the same visual size
though differing extraordinarily in their visible size, for it is the
moon that we see, not its visual sign. All this we frequently confirm in our daily lives. I can deduce that the meanings we attach
to the words of visual language may be different for a grammarian
who minutely and constantly examines them, as the experience
reported by Miss Lucia Ronchi confirms.

Moreover, the same resilience or adaptability of the Linguistic
Theory is manifested by its use of an unusual property specified
in the model: the words of visual language are often ambiguous,
and sometimes their meanings are diametrically opposite to their
customary meanings. To choose a different example from the
moon illusion, when we look into a plane mirror the meanings
left and *right* are reversed while *up* and *down* remain the same.
But most of us have long since managed to avoid being taken in
by the ambiguous language of the plane mirror. How easily can my
theory accommodate these facts by applying the resources of the
model! We avoid ambiguity only through awareness of the context. Only after long experience are we able to avoid being victimized by the metaphors and other ambiguities of ordinary and
visual language so that we can tell that such things as General
Contentment wear no stars and sticks crooked in the context of
water have no corners.

Finally, if, in the present test-case, faintness is accepted as the
main cue, the relative economy of the Linguistic Theory is dis-
played. One traditional solution is that the horizontal moon looks
bigger because it looks farther away; it looks farther away because
the sky looks or is imagined flattened; and the sky looks or is
imagined flattened because of intervening objects in the landscape.
In my solution, the moon looks bigger *and* farther away because
it looks fainter.

I have already concluded that this test, being one in which the
Geometrical Theory capitulates before the issue is joined, plainly
shows that because there are phenomena in vision lying forever
beyond the reach of geometry, a new theory was needed. From
the argument of this section I am now able to conclude that the
Linguistic Theory not only gives a satisfying solution to the problem but leaves room for further testing.

4. The Inverted Retinal Image

I have left this test-case till last, yet it is the most significant in the series. It predominantly elicits the virtues of one theory and exposes the defects of the other. The solution of the problem requires ingredients that, it seems, are uniquely owned by the language metaphor. Yet these ingredients are commonplaces of language. So commonplace are they that we rarely take notice of them. I am able to reach an obvious yet amazing solution. In addition, the metaphor offers more than is demanded of it. It suggests a correction to a mistaken conception of the mind—a suggestion that I shall barely mention in this book—and it brings some clarity into a concept which, it seems, is the private property of geometry, namely, the optical image, a subject still obscure and controversial.

Apart from all this, the solution to the present problem is an exercise in the use of metaphor extended by stages, of which there are three. In the first I apply the metaphor to the problem as it was originally set, and solve it. But in doing so I use only part of the "system of implications" of the metaphor or, if preferred, only some of the features specified in the model, meanwhile holding the others in reserve, as it were. This involves sustaining the make-believe that the geometrical solution is partially correct. In the second and third I extend the metaphor further to displace some implications of the geometrical solution.

The problem is this. How do we see things erect although the images of them are inverted on our retina? The discovery of the retinal image proved both a boon and an embarrassment to optical theorists, a boon because now the "mechanism" of the eye in vision was completely solved by the "discovery" that the eye is literally a camera with a converging lens and a screen that records pictures of the external world, and an embarrassment because the pictures, being upside-down, needed heroic measures to explain how we see things right-side-up.

How was the problem solved? Common to all the solutions were three ingredients: (1) When we see objects in space there are images of them painted on our retinas exactly like the objects in shape and color except only that they are diminished and inverted; (2) we are able to use these retinal images or the movements by

which they are formed as instruments in vision; and (3) we see
by natural geometry; that is, we compute or deduce what we see,
but without noticing that we do so, and specifically without noticing the inversion of the image. In order to illustrate (3) Descartes
used the device of a blind man who "sees with his hands" by holding two crossed sticks—*crossed* to re-invert the inversion of the
image, and *sticks* because the rays of light are straight (FIG. 11).

Figure 11

Just as the blind man can "see" the positions of objects by feeling
with his sticks, "without in any way having to know or think of
the inverted position of his hands," so does the soul see the
real positions of objects by "feeling" the impulses of the rays of
light without having to know their inverted positions.[19] Molyneux's account, already noted, was similar. The mind "takes no
notice" of the inversion but "hunts back" along the rays.

Having accepted (1) and (2), it is hard to conceive what other
solution the geometrical theorists could devise than one that
involves making a deductive inference and taking no notice of the
inversion. I have already given the aetiology of such a solution.
It is plainly a case of exporting without awareness an item of procedure to the process of nature. From one point of view, it is a case
of confusing the geometrical construction of the position of the
object from the position of the real image (all of which can be
done on paper) with how we see. It is, therefore, a case of sort-
crossing, and since this arbitrary sorting is taken to be the correct ontology of vision, of sort-trespassing.

19. *Dioptrics*, VI. The diagram copies Descartes'.

But from my point of view, since both the sort-crossing and the make-believe are now clearly apparent, it becomes a case of using a metaphor and of extending the metaphor to an extraordinary degree. We are to make believe first, that we deduce what we see; secondly, that we make the deduction without noticing that we make it; and thirdly, that we make it successfully although the premise is the direct contrary of the conclusion. I cannot reject the geometrical solution for the reason that it is based on such make-believe, for, if I do that, I must also reject my own. The test will be a test between two theories based upon two different metaphors, and the choice between the two theories will amount largely to asking: "Which is the better make-believe?"

Let me, at this stage, grant certain things for the sake of testing the two theories in their answers to the main problem. Let me, therefore, grant, but only for the time being, ingredients (1) and (2) of the Geometrical Theory. That is, let me sustain the make-believe that retinal images are colored pictures painted upside-down on our retinas and that we use them or their corresponding movements as instruments to see with. In which case I too have to try to explain how we see things erect although our retinal images are inverted. In order to do so let me discard ingredient (3) by dropping the pretence that we see by geometry. Instead, let me pretend that we see by language. Once I do this, I find no difficulty in bridging the gap between inverted retinal images and the erect things we see. For I displace the premise-conclusion relation and replace it by the relation between signs and things signified, the inverted images becoming the former and the erect objects the latter. In which case it is as if all our books have to be read upside-down. The question becomes: Would we give the marks the same meanings they now have, or would they acquire opposite meanings? Would the marks INVERTED come to mean *erect*, and would RIGHT come to mean *left?* The answer is fairly obvious. From the properties specified in our model I know that when we read or listen we overlook the marks or sounds themselves and pass on to the meaning; that though we may hear the sounds RAKS, it is rocks that we think of, not RAKS; that the erection or inversion of the marks INVERSION is not what we attend to at all but rather the posture of the object signified; and that a little experience would enable us to bridge the gap between the marks SⱢVƆ and cats as readily as we now bridge it between CATS and cats. From all this I can deduce that there is no difficulty in seeing things erect although their images are inverted on our retinas. For, being signs, the erection or inversion of the retinal images is quite irrelevant to what they signify.

Which theory offers the better solution? Can I devise an experiment which may disconfirm one theory? It would be nice if I could find someone already equipped with erect retinal images. Since this is difficult, however, all that I need to do is to produce this predicament in myself artificially by binding inverting lenses to my eyes, thus re-inverting the inversion of the retinal images. What would the consequences be? Would the world forever look topsy-turvy?

If so, it would confirm the geometrical solution. For, according to the Geometrical Theory, it follows that the inversion of the retinal image is necessary to erect vision, and therefore, that to someone with erect retinal images, the world would look forever upside-down. This is so because, according to the theory, the relations *up* and *down, left* and *right, erect* and *inverted,* are things fixed and immutable, like Newton's absolute time, place, and motion, and because these relations are one and the same for touch and sight. Accordingly, what is "received" on the bottom and top of the retina must be related to what is seen as up and down, respectively. Descartes' device of the blind man with crossed sticks is a good illustration of this. It is impossible for such a man to feel with the bottom hand the movements originating from the bottom. Why? Because the movements do not come from the bottom. They come from the top.

On the other hand, it would mean the collapse of my theory. For, in accordance with my theory, I can deduce that inverting spectacles, bound to my eyes would impede, but only for a few days, my understanding of the words of either an ordinary or visual language. The marks ꝆƎⱢᴚƎ∧NI would not mean *erect;* I should not point downwards at the moon; and the Eiffel Tower would not become the Eiffel Well. Everything should before long look erect or in its proper posture. But this, so it happens, is the case, as anyone can confirm. The experiment was actually per-

formed on himself by G. M. Stratton who, after wearing an inverting lens night and day for a period of eight days, found that he
reacquired "erect" vision.[20] At first, as he reports, he suffered ver-
tigo and nearly vomited. But slowly, day by day, as the old associations were unlearned and the new ones established, the world
ceased to look topsy-turvy. When he took off the lens, however,
he had to repeat the same painful process.

It seems that the experiment just described is crucial between
the two theories. Once I grant that retinal images are inverted and
that they or the movements by which they are formed are instruments or means, the problem of choosing between the two theories
reduces to this: Which theory can more economically accommodate both the inversion of the image and the erection of the
object seen? Now both theories use, as it seems they must, the feature of "taking no notice" of the inversion. The problem then
becomes: Which theory can better fit in this feature? Two different models are used. In one, from a given premise, I deduce or
compute a conclusion or answer. But since the premise is the
contrary of the conclusion, I have to ignore it. This is queer. So
an auxiliary model, the blind man with crossed sticks, is used.
This enables the theory to fit in the inversion and the "taking no
notice," but not the reinversion of the experiment. In the other
model a sign suggests to us a thing signified. Its position is irrelevant to what it signifies, and we overlook it. The Linguistic
Theory can fit in all the facts without using an auxiliary model.
To sum up, while the Geometrical Theory solves part of the
problem extravagantly, the Linguistic Theory solves the whole
problem most economically.

What precedes solves, so it seems to me, the problem as it was
set, and solves it with greater ease and expedition than any other
theory known to me. But this achievement does not exhaust the
resources of my model. Some things have been left out because
they were not needed. These I shall now add. I move, then, to the
second stage in the argument, and to the destruction of certain
implications of the geometrical solution.

20. G. M. Stratton, "Vision without Inversion of the Retinal Image," *Psychological Review*, Vol. IV, No. 5 (Sept. 1897).

Let me continue to keep ingredient (1) of the Geometrical Theory. That is, let me grant that we own retinal images considered as colored pictures. But let me proceed to do without ingredient (2), which is that our retinal images or the movements accompanying them are instruments that we use to see with. There were two important versions within this solution. The first was described by Molyneux: "The rays . . . determine . . . on the retina, there painting distinctly the vivid representation of the object, which representation is there perceived by the sensitive soul."[21] Entailed by this view is the presence of a ghost lurking behind the eyes who *"sees"* the colored pictures and makes inferences to external objects. The second, the Cartesian, was a correction of the first: we must not think that we look at the pictures "as though we had another pair of eyes to see them inside our brain"; rather, we must hold that "the movements by which the pictures are formed act directly on our soul."[22] Entailed by this view, then, is the presence of a different ghost who does not *"see"* the pictures but who *"feels"* the corresponding movements and makes inferences to external objects.

Now I have concluded that even if I accept either of these solutions the preliminary problem of the inversion is easily solved by my theory. But there is no need to do so. Let me sustain the make-believe that we see by language, applying the metaphor this time beyond the sign relation to the nature of the signs themselves. In which case our retinal images cannot be signs. Why?

I know at once that although we normally overlook the signs of a language when we read or listen, they are noticeable. Indeed, they must be in order to function as signs. Moreover, I know at once that although signs may have situations among themselves, this situation is of a different type from the situations of the things they signify. The marks INVERTED do not have to be inverted in order to be used to mean *inverted,* and the sounds hαʏ do not have to be pronounced in a high key in order to be used to mean *high.* Certainly men could have used ᗡƎⱢᴚƎɅͶI and raised their voices if they had been fanatical devotees of the picture cult of meaning. But it does not make sense to say that these marks

21. *New Dioptrics,* pp. 104–05.
22. *Dioptrics,* VI.

and sounds must have the same situation as, or a situation relative to, the situations of the things they signify. Accordingly, I
can deduce that the situations of visual signs belong to a type different from that of the things they signify; that colors are not high
or low, straight or crooked, erect or inverted, in relation to what
we perceive by their means; and that if we confuse them we might
just as well say: " 'The Eiffel Tower' and the Eiffel Tower are
erect," or " 'Paris' is to the left of Paris."

But our own retinal images or their corresponding movements
are not noticeable, in the way that looks and feels are. They are,
moreover, inverted in relation to the objects we see in space. It
follows, therefore, that our own retinal images or their corresponding movements cannot function as signs in the Linguistic
Theory. In which case I can easily abstain from the implication
of the presence of a ghost in the head who *"sees"* the retinal
images or *"feels"* the corresponding movements.

Which, then, is the better picture: that of a ghost behind the
eye who looks at a screen or feels the movements and makes inferences to the external world, or that of an auditor or reader who
understands and acts upon the signs of a language? Which produces the better theory? It seems that there is no difficulty in the
whole problem that my theory cannot solve. The problem is
solved with great ease whether or not I make the queer assumption that I can use as instruments events occurring inside my head.
This is not the case with the Geometrical Theory which, though
it provides a solution, does so with difficulty by using many ingredients that seem remote from the facts as they generally appear
to us.

I come now to the final stage of the argument and to the clarification of ingredient (1) of the geometrical solution, which is
that the retinal image is a colored picture. As we have seen, the
optical theorists first *inferred* the existence of the retinal image
through their knowledge of lenses and of some anatomy. But they
thought they had *discovered* its existence, for they could actually see a colored picture on the back of another eye. This picture
was the retinal image, its existence having been maintained in
both versions I have been considering, one holding that we

"see" our own image, the other that we *"feel"* its corresponding movements. Its existence, it seems, was confirmation that it is the immediate object of sight. What else is its purpose?

Now although the retinal image was discovered in the early seventeenth century, and although it immediately occupied a central place in all accounts of the mechanism of the eye in vision, much confusion still surrounds the question of its nature. The same conclusion can be drawn in regard to the real image—of which the retinal image is a special case—and by parity of reasoning, to any optical image, including the virtual image. Is the optical image (a) a colored picture that can be seen on a screen, or (b) a mathematical entity constructed out of lines and angles, and therefore invisible, or (c) an equally invisible physical event consisting of inferred energy effects?

Let me suppose that the optical image is a colored picture. This, it seems, is by no means an irrational supposition, for it seems that we can see a virtual image on every occasion that we look through a magnifying glass or diverging lens or into a plane mirror, and that we can see a real image on every occasion that we look at a cinema screen. This is the way writers on optics have written since the time of Kepler, except that Descartes, Barrow, Newton, and others, following Kepler's usage, called the virtual image "the image" and the real image "the picture." Thus Descartes said that "you will see a picture," while two recent writers, F. A. Jenkins and H. E. White, in their *Fundamentals of Optics,* say that when one looks into an aquarium, "one is seeing virtual images which are not in the true position of the objects."[23] But if I suppose that we see the optical image, strange consequences follow.

One is that countless numbers of things seen through lenses and in mirrors do not exist. This follows from consideration of the Barrovian Case. Barrow said: "As a matter of fact, by the word *'image'* I mean nothing but the place from which many rays seem to diverge."[24] The optical writers of the seventeenth century thought that this rule worked or should work to determine the place of the thing seen. They thought that the diver-

23. *Fundamentals of Optics* (New York: McGraw-Hill, 1950), p. 21.
24. *Eighteen Lectures,* Lect. 3. 16, my translation.

gence of the visual rays from the place of the image of an object was the cause of its appearing in that place, and, therefore, the same cause must hold good in vision after reflection and refraction. But the Barrovian Case shows that the figure seen cannot be the real image because the rays meet behind the head. It shows that it cannot be the virtual image because in a virtual image, in the words of F. A. Jenkins and H. E. White, "the rays from a given point on the object do not actually come together at the corresponding point in the image; instead they must be projected backward to find this point," where, to the observer's eye "these rays appear to be coming from."[25] This projection backward is impossible here because there is no point where the imaginary rays can be projected to meet. Yet the figure is seen. To conclude that it does not exist is absurd. It is better to conclude that the figure seen is one thing and that the optical image is another. What, then, is the optical image if it is not a thing seen? That case known as the virtual image, in which nothing physical is supposed to pass along the rays but which is defined as the intersection of these virtual rays, is clearly a mathematical entity with the same ontological status as the equator or a Euclidean point.

What, however, are the consequences if I suppose that the other case of the optical image known as the real image is a colored picture, that is, is something visible? Consider that special case of the real image known as the retinal image. Descartes and others saw a colored picture on the retina of another eye and could not doubt that a similar picture was present on their own retinas. Now if I suppose that the real image is a colored picture, then I must conclude that our retinal images are colored. In which case a strange duplication of colored objects follows, namely, colors on the external object and colors on the retina. By analogy I may infer a similar duplication of sounds, smells, and tastes, and, therefore, that ear, nose, and mouth images are audible, odorous and tastable. If this is so, it is hard to resist the further inference to additional sense organs inside my head.

All this is avoided if I treat the real image as entirely distinct from the figure seen. Strangely enough, the seventeenth-century optical writers had available the means for avoiding this confu-

25. *Fundamentals of Optics*, p. 22.

sion. They forgot their own distinction between *lumen* and *lux,* that is, between invisible physical light made up of corpuscles, and light considered as color that can be seen. The rays of light are not colored. They are, as Newton said, "color-making." Yet, at the same time, like Descartes and others, he wanted to say both that we can "see the pictures" and that "the pictures are transmitted into the inside of our head." A theorist may see colors on another retina just as he may smell and taste events in olfactory and gustatory areas other than his own. He may suppose, in line with his theory, that invisible physical events corresponding to external objects are transported along nerve pathways to the brain. But to suppose further that retinal images are colored pictures is analogous to supposing that there are smells and tastes inside our heads.

Using the linguistic model, I know that the pictures we see on the backs of other eyes or on a cinema screen are analogues not of words but of what words signify. They have position, size, shape, and distance in relation to all the other things we see. These pictures, therefore, are not signs for their owner, nor is their status as pictures (that is, their resemblance to the object) essential to their being members of the causal chain supposed to occur in a physical explanation of seeing. From the Barrovian Case I know that retinal images, being cases of real images, are distinct from pictures or other things seen in space, because real images can be supposed to exist behind heads, when no screens are present, or independently of the presence of eyes. Moreover, for those theorists who account for vision in a physical way, it may be useful to suppose the existence of physical *light* or *lumen* which affects cells, emulsions, and eyes, and which is refracted or reflected by optical systems. In which case, consistent with this supposition, the retinal image may be defined as a collection of invisible physical events occurring on the retina.[26]

This brings some clarity to the concept of the retinal image and, by the same token, to the concept of the real image. It also brings

26. Cf. *Optics: The Science of Vision,* sec. 267, where Professor Ronchi defines "retinal image" to mean "distribution of energy effects on a layer of the retina." For a remarkable clarification of the subject of the optical image, see his whole chapter, "The Optical Image."

clarity to the wider concept of the optical image by providing the fundamental distinctions required for its definition. In treating these concepts, optical writers must decide upon the obvious preliminary matter of whether they are treating the image as either (a) the thing that is seen; or (b) the invisible geometrical device; or (c) the equally invisible physical event such as supposed energy effects.

Conclusion

1. Language or the Camera

THE GEOMETRICAL THEORY I have presented is the core of all subsequent representative, copy, or picture theories of perception. The most widely held solution to the problem of perception among scientists and philosophers from the time of Kepler and Descartes has been the Representative or the Copy Theory. This solution preserves Aristotle's distinction between what we are directly aware of and what we infer or judge, between sensing and perceiving. In the words of Locke, "the mind knows not things immediately, but by the intervention of ideas it has of them."[1] Because our sense-data represent or picture their physical causes we can make inferences to them.

Some such solution, it seems, is demanded if we take seriously the cases of illusion, hallucination, and relativity, and the great optical, anatomical, and other physical discoveries of the seventeenth century. The Representative Theory was designed to accommodate them. But there have been different versions of this theory. Characteristic of their differences are the different models used. The main model is found in the Geometrical Theory of Vision.

This model is the camera. As we have seen, hundreds of years were needed to perfect it. Ibn al-Haitham was the first I know of, although he said he was not, to use the camera to illustrate the eye. "What is true of the camera," he said, "is true of the eye." It was left to Kepler, the inventor of the first portable camera obscura, to perfect the model and to apply its features point by point to the eye. The eye is a camera, a machine for taking pictures of the external world. It is equipped with an aperture, a light-sensitive material, a converging lens, a focussing mechanism,

1. *Essay,* IV. iv. 3.

and a screen on which the pictures are received. All this, represented by Rules (1) and (2) of Kepler's solution, was beautifully described by Newton in his famous Axiom VII. He used the phrase "in like manner" to describe the relation between the camera and the eye. This, however, was only the first half of the application of the model to vision. It provided the universally accepted solution to only half of the whole problem that had been set by the Greeks, namely, the problem of *the dioptrics of the eye in vision*.

The second half of the application of the model to vision made possible, so it was thought, the solution to the whole problem. We see by using the photographs on the back of the camera. We do not take the picture. This is done for us merely by opening the aperture in the light. The mind, therefore, is passive in receiving these visual data. But then vision is accomplished by working back from the photograph of the object to the object itself, or if a lens or mirror is interposed, to the virtual image. This working back involves inference grounded upon the resemblance and necessary connection holding between the photograph and the object. We perform the inference much after the same manner as we make a graphical construction of the object in geometrical optics when we are given the real image and the focal length of the lens. All this, represented by Rules (4) and (5) of Kepler's solution, was implied by Newton in his Axiom VIII. Consider how this accommodates visual illusions. Mirror and lens' illusions are virtual images. The solution explains how we see the apparent position of the shark in the aquarium and the appearance of the bent stick partially submerged in water. The extended application of the camera model thus provided a solution to the second half of the whole problem, the problem of how we see objects. *The mind does not see physical objects immediately, but by inference from the images or photographs it has of them.*

Because of the illumination it seemed to give, the camera model captured the imagination of scientists and philosophers. They extended its application. Kepler applied it only to the eye, but his pupils Descartes and Newton, and the latter's "under-laborer," Locke, created the popular view of the mind by applying it to the whole understanding:

For methinks the understanding is not much unlike a [camera or] closet wholly shut from light, with only some little openings left, to let in external visible resemblances or ideas of things without. Would the pictures coming into such a dark room but stay there, and lie so orderly as to be found upon occasion, it would very much resemble the understanding of a man.[2]

By this extended application, Locke was describing the *dioptrics, as it were, of the mind in perception.* He and his colleagues proceeded to represent the facts about the mind in the idioms appropriate to cameras. Just as there are images on the back of the camera so there are *ideas in the mind.* Just as there are obscure and confused, clear and distinct images so there are the *obscure and confused, clear and distinct* ideas. Just as we inspect the images on the screen, so we *introspect ideas* in the mind. Just as images are reflected light, so there are *ideas of reflection.* These idioms, some of which are still fashionable, refer to central concepts in seventeenth- and eighteenth-century epistemology.

But just as we are passive in receiving the photograph, the camera doing the job for us, so are we passive in receiving the ideas of external objects. The "understanding" (Saxon) or "substance" (Latin) merely receives its contents. It would be a mistake to think, however, that since the mind is wholly passive in receiving its ideas, it is passive in perceiving objects by means of them.[3] "As the mind," Locke says, "is wholly passive in the reception of all its simple ideas, so it exerts several acts of its own."[4] These acts include combining, comparing, and separating or abstracting. They also include working back from the ideas to their causes. *The mind, therefore, knows physical objects as a result of inference from ideas grounded on conformity.*

What are the defects of the camera model? It would be vain to say that the eye is not really a camera which takes pictures of ob-

---

2. *Essay,* II. xi. 17.

3. Cf. Stuart Hampshire, *Thought and Action* (London: Chatto and Windus, 1959), p. 47: "The deepest mistake in empiricist theories of perception descending from Berkeley and Hume has been the representation of human beings as passive observers receiving impressions from 'outside' of the mind."

4. *Essay,* II. xii. 1.

jects, or that the mind is not really an understanding or substance which receives into it ideas of the external world. I might just as well say that people who carry all their eggs in one basket do not really have any eggs or any basket. But its defects outweigh its merits. It is interesting to notice the heroic measures adopted to make it work.

First, the interpretation of the photograph is not built into the camera. The model is marvelously equipped for illustrating the taking of the picture but not for illustrating the interpretation of it. It sheds light on the passive aspect of the mind in vision, but it can shed no light upon the active aspect of vision. It can suggest an account of the reception of the visual data but not one of the perception of the mind-facta. Here the limits of the camera model and of geometrical optics become apparent. They offer a dioptrics of the eye but not, in one good sense of the word, an optics. With Kepler, geometrical optics acquired a new purpose, the making of optical appliances. It is true that craftsmen before his time, ignorant of theory, knew how to make spectacles, microscopes, and telescopes, just as Captain Cook knew how to prevent scurvy by eating sauerkraut without knowing why it did prevent it. But Kepler provided the theory, improved the product, and thereby established the ulterior purpose of geometrical optics. In order to produce a theory of vision, however, he had to put a ghost inside the camera just as he had to get inside his own camera obscura. This allowed for the interpretation of the photograph. Now it may be of value to make believe that such a camera exists. But, as I showed in the last section, even if we do, it is far better to treat the photographs as signs that convey meanings rather than as pictures of some original.

Similarly, if the camera is used to illustrate the whole understanding, a corresponding defect is elicited. It can illustrate the "dioptrics" of the mind, that is, the mind as passive substance or understanding which receives, supports, holds together, and owns its ideas. But it cannot illustrate the "optics" of the mind, that is, the mind as actor which combines, compares, and interprets its ideas. Locke said that the contents of the understanding are "to be found," but the understanding does not do the finding. Another active soul is needed, therefore, for the interpretation of the contents of the understanding. Both concepts, the mind as substance or owner of *its* states and the mind as actor, are still present in the ordinary concept of person. Now even if I accept this double make-believe of ideas *in* the mind and of a ghost who interprets them, it is far better, as we have seen, to treat these ideas as signs instead of as duplicates.

Secondly, how do we know that the photographs in the camera are copies of some originals when we never find the latter? How do we know that our sense-data represent physical objects when we never make the comparison? The optical theorists thought that they could compare the photograph with the original. In this they were deluded, for, according to their own theory, they could contemplate only their own images on the screens of their own cameras and compare them one with another. Early subscribers to the Theory of Representative Perception thought that they could compare sense-data with physical objects. How otherwise could they know that the former copied the latter? In this they were deluded, for, according to their own theory, they were confined to the contemplation of sense-data. Growing awareness of this difficulty elicited elaborate proofs of the existence of the external world from philosophers. Subscription to the Linguistic Theory, however, renders such proofs redundant. Signs do not copy or picture undiscoverable causes; these signs are tied to the world, for they convey meanings which, in the form of hypotheses, we can reject or retain; and the questions we ask of the external world, such as "What does this mean?" and "Is this a dagger which I see before me?" are answerable within the rich context of further experience.

Finally, in consequence of these defects, geometrical and other representative theories lack an adequate solution to the problem of illusion, the problem for whose solution they were designed. Thus I might have said that they lack an adequate solution to the problem of perception, for it was cases of illusion that created this problem. They cannot offer a solution unless they first place a ghost inside the camera or the understanding. But this is not enough, for the size, shape, and color of the photograph plus an innate geometrical ability are all the ghost has to go on. He can make a good inference to the crooked oar, but he cannot *see that* it is straight. He can make a good inference to the yellow cup, but he cannot *see that* it is white. He can make a good inference to the apparent nearness of the object in the Barrovian Case, but he cannot *see that* it is really far away. He cannot *see that* the horizontal moon is no bigger than the meridional one although the size of the photograph he contemplates hardly varies. He can see the inverted pictures on the camera screen, but he cannot *tell by sight* that things are erect unless, heroically, he ignores the inversion.

In many cases when we see things through lenses and in mirrors he should see nothing at all because there is no virtual image. He cannot tell by sight which is bigger, his thumb, or the Eiffel Tower, because sometimes the former is much bigger in his visual field, at others the latter. He must even be fooled by the images seen in the plane mirror. In all these cases he must be fooled by the *ambiguities* of vision in much the same way as children and mechanical brains are fooled by *metaphors*. He must be fooled unless he is a ghost who can interpret what he sees from the *information* given to him, a ghost with a memory who can tell and often mis-tell what things *signify,* and who, aware of *contexts,* can sometimes see through the *ambiguities* and *double-talk* of vision. In short he must be fooled unless he has learned to *understand* the *words* of the *language* of vision.[5]

2. *Language or the Machine*

But I have had another purpose ulterior to that of offering the Linguistic Theory as a rival to the Geometrical Theory in the concrete problem of vision. In Part One of this book I began to insinuate the notion that the linguistic metaphor might be used with success to illuminate other areas. This could produce no con-

5. Perhaps aware of the defects of the camera model, followers of the Representative Theory have devised other models. Among these are symbolical representative devices, such as the telephone exchange and the telegraph. According to the latter, messages, coming from the physical object in code, are decoded by the brain into a representation of the physical object. There are also non-symbolical representative devices such as the cinema and television. According to the latter, we see physical objects indirectly by means of the images on the screen. It seems to me that these models share most of the defects I have ascribed to the camera. For example, how do we tell by sight that the stick in the studio is crooked when all we are given on the TV screen in the living room is a straight one?

viction until it had been put to work in an actual case. Fortified
by its successful performance in the case of vision I can now adven-
ture with it in other areas. Once more, however, in these other
areas, its chief competitor turns out to be the ubiquitous geometrical model or that extension of it that should be known as the
model of the machine but which in fact is known as science, for
it is confused with what it models. In Part One I was able to extract the features of the geometrical model as it was understood
and used by Descartes. For him, it was the same as the mechanical
model. It differed in some respects from the geometrical model as
applied to vision in Part Three. But the two versions shared the
use of the deductive procedure and geometrical symbols, and both
exported the deductive relation to the process—to the process of
nature in general in one case, and to the process of vision in particular in the other. Thus I can say, either that the geometrical
and the machine models are the same thing, or, more appropriately, that the former becomes the latter merely by adding to the
system of solid geometry the primitive notions of motion, in the
case of Descartes, or motion plus force, in the case of Newton.

Let me therefore treat the events in nature *as if* they compose
a language, and let me compare this model with the other model
that still dominates all others used in science, the machine. Fulfilling Descartes' dream, it dominates all other illustrations of the
physical world. Fulfilling his dream, it begins to dominate the
world of living things, for in biology mechanism has elbowed out
vitalism and neo-vitalism, and is elbowing out emergence. The
time is even imminent when a neo-Cartesian may exclaim: "Give
me extension and motion and I will construct life." But shock-
ing to his dream, it also begins to dominate earlier rival illustrations of the human mind. All this is so in spite of the meager
opposition offered by the theologians, a few poets, and fewer
philosophers who, in general, have been victimized by their own
metaphors to the same degree as their opponents, They have opposed one metaphysics by another. Now that we know that we deal
in metaphors, however, I shall try to keep the competition on that
level.

Both the machine model and the language model start with the
same facts: phenomena which strike on the senses co-exist or suc-

ceed one another. On the old model these are interpreted as effects of efficient causes, hidden in the works of the machine, moving of necessity. The movements of the giant clockwork "depend on parts so small that they utterly elude our senses." [6] On the new model these are interpreted as the signs of a language which suggest things signified without any necessity, and the signs must be noticeable.

On the old model, just as a mechanic "with experience of machinery . . . can readily form a conjecture about the way its unseen parts are fashioned," so the mechanist seeks "to investigate the insensible causes and particles underlying" the mighty machine or giant clockwork of nature.[7] He thus discovers the works or the "go of it," and calls them the laws of nature. To explain the works, the mechanist shows what effects necessarily proceed from these causes. On the new model, just as every language has a grammar, "there is a certain analogy, constancy, and uniformity in the phenomena or appearances of nature, which are a foundation for general rules: and these are a grammar for the understanding of nature." These rules we learn by experience, and the grammarian's task is to reduce statements about phenomena to these general rules by making believe that these rules are principles or true statements from which other statements are deduced (S, 252).

On the old model there are different degrees of understanding the machine and of skill in using it. One can learn to use a machine without knowing how it works, but the expert mechanic who knows the laws of its operation can better foretell its motions and pronounce on its application. Similarly, the language of nature is interpreted and used with different degrees of skill. There are two ways of learning it: "either by rule or by practice; a man may be well read in the language of nature without understanding the grammar of it." Thus through frequent practice, he can predict and act accordingly without being able "to say by what rule a thing is so or so." But the best grammarian not only foretells the motions of the planets but reduces these motions to rule.

Extending the mechanical metaphor to apply it beyond the physical world, as Newton did, "we come to the very first cause,

6. *Principles,* IV. 203.
7. Ibid.

which certainly is not mechanical."[8] The machine requires an inventor and starter with knowledge of the machinery including geometry, optics, etc., and this mechanic is no part of his machine. But the primary laws of motion discovered by the mechanist are such that, once made and moved, the machine is *auto*motive. The inventor is thus a remote mechanic or an impersonal and distant deity, and mechanists who first extend their metaphor this far and then take it literally are mechanists who are also deists. On the other hand, if nature is a language whose terms are natural signs delivered to us according to rules, and we extend the metaphor to apply supernaturally, we come to the author of this language. The signs imply a will, and their grammar rules a will "conducted and applied by intellect." Unlike the great mechanic, however, the great author is constantly required. He therefore *speaks* rather than *writes,* although we may hear what he says with our eyes. The author is thus immediately present, a personal God in *rapport* with his audience, and grammarians or listeners who first extend their metaphor thus far and then take it literally are theists.

Why choose one model rather than the other? Both are seen as two different ways of speaking about the same facts or as two different pictures of the same subject. Let us suppose that the corresponding laws of the machine and the rules of grammar, although differently stated, are equivalent, and that to "undress" both interpretations is to reveal the same formal deductive system. There are three apparently different points of view from which the question may be regarded: as a *conflict* between science and metaphysics, mechanics being science and the language of nature being a metaphysical theory; as a *conflict* between two opposing metaphysical theories, scientific theories being found to presuppose metaphysics; or finally, as a *choice* between different models, either of which can be used to illustrate or explain. The first point of view—and the one that would be most fashionable because there are more mechanists than grammarians, and more mechanists than those who take the language metaphor literally— is naïve, like the apparent conflict between science and the humanities. And the second reduces to the third.

8. *Opticks,* Query 28.

There are well-known tests that may be used for choosing between competing scientific theories. Since the metaphors under consideration are used in science, the same tests will apply. Now that we know that we are choosing between metaphors, however, some additional tests reveal themselves.

Newton and Descartes drew their metaphors from three sources: Euclid's geometry, persons, and machines. Now there is no harm in using metaphors or in mixing them. In either case the price is vigilance, but in the latter the price is increased. Metaphor within metaphor is especially dangerous, for, as H. W. Fowler points out, combination is one thing, and confusion is another. Nothing, he says, can save the time-honored example "I smell a rat: I see him hovering in the air. . . . I will nip him in the bud."[9] The case is much the same when the metaphors become models, only the idioms are different. Models combined with others, like the wave and the corpuscle, are allies, while models within models are auxiliaries. The combination can result in redundancy, and, sometimes, confusion. The former, in the language of models, issues in lack of economy or simplicity; in the language of metaphysics, it multiplies entities beyond necessity. Now Descartes and Newton mixed their metaphors to produce both redundancy and confusion. Their world-machine, once invented by the mechanic, was *auto*motive, but the mechanic, although no longer required, continued to preside over it. In addition, however (again unnecessarily, since their mechanics could have been merely kinematics), they added forces to every movement of the machine. Thus nature is a machine with many ghosts inside it and one ghost outside it: many occult forces and one supreme occult force. This double redundancy issued in confusion also, for the many forces are corporeal, and the one force incorporeal, just as though we were to say: "The great mechanic forced forces to move all the parts of his giant clockwork." The modern followers of Descartes and Newton, who do not extend their metaphor to apply supernaturally, avoid this confusion, but retain the redundancy. For them there are gremlins in every self-starter, automatic transmission, and generator of every automobile. On the other hand, the language metaphor avoids both the

9. *The King's English* (Oxford, 1906; 3rd ed., 1930), pp. 212–13.

confusion and the redundancy. This is so because there is no mixture of metaphors in the series: language, signs, things signified, rules of grammar, and (if the metaphor is to be extended as far as Descartes and Newton did theirs) author.

There is a well-known rule for judging illustrations: we illustrate the unknown with the known. The means chosen to illustrate must be familiar. We are all familiar with force or power in ourselves, and most of us use machines at various times. We are in general more interested in machines than in language, and are fascinated by them. We are fascinated because they resemble us. It is *as if* they can make and do things without our help. It is *as if* they contain forces or powers which produce their movements. This factor partly accounts for the early success of the machine metaphor and its continued success. Moreover, machines were once rare events. Things that rarely happen strike. Nevertheless, though many use them, few know how machines work. We are barely familiar with their parts, less with the laws of their operation. They are a mystery to all but the expert. On the other hand, all of us use a language constantly. Many more know better how to use a language than how to use a machine. It has, indeed, grown so familiar that we are blinded by excess of light. Again, however, we can use it without knowing the rules of its grammar. There are experts here too. Accordingly, from the standpoint of familiarity with laws or rules, the choice may be difficult, but from the standpoint of familiarity with use, the choice is easy.

An obvious way to choose between rival metaphors, like the way in which we choose between different portraits of the same subject, is by their degree of likeness to the thing illustrated. We can ask this in general independently of whether, as scientific theories, one saves more appearances than the other. We know that the world is not a machine with ghosts in it, nor is it a language. But which is it more like? What this amounts to is this. Which is closer to the natural process as we found it to be after stripping it of the metaphysical disguises put on it by Newton and Descartes? First, is the relation that we observe to obtain between events more like that of cause-effect, or more like that of sign-thing signified?

The surface-level answer to this question, already discussed, is that the relations we find in the process are those of co-existence and succession, and these are nothing but the immediate ancestors of the relation between a sign and what it signifies. Efficient causes in the natural process have yet to be found. Moreover, we have seen that the relation of necessary connection, far from holding in the giant clockwork of nature, does not even hold in any actual clockwork. It is true that the actual clockwork goes "tick-tock, tick-tock," with monotonous regularity, but this and the corresponding movements are all we find. The delusion that, in addition, the movements are necessarily connected, is explained as a metaphysical addition to process from procedure. Secondly, and for the same reasons, the laws found to obtain in the course of nature are less like causal laws than like rules of grammar. The former connote production of existence, necessity, and universality, the latter regularities that have exceptions.

But the deeper-level answer to this question is in part that we cannot answer the question, because we can never be sure what the facts are. I have already made the point that a new metaphor changes our attitudes to the facts. Once we see the world from the point of view of one metaphor, the face of it is changed. But when do we become aware of all the metaphors so that at last we know we are confronted with the unmade-up face of the truth? Most of the metaphors come disguised. This, it seems, we have to accept. The other part of the answer is that a good metaphor, like a good portrait, does not hold a mirror up to the face of nature but vividly illustrates some features of it and neglects others. It is probably a better metaphor even if there are some differences which may throw into relief the startling likenesses. From these factors it is hard to choose between our rival metaphors.

The mechanical metaphor pictures two worlds corresponding to the surface of a clock and its internal constitution. The world of causes is forever hidden from view, but we can make inferences to it in virtue of the *structural* resemblance existing between causes and effects. The effects are only appearances or ideas: some of them copy the primary qualities of bodies, while some only correspond to the powers of primary qualities, namely, the secondary qualities. But this picture of the world that might be called "The Two-World Picture" creates difficulties. It seems that if one thing is said to resemble another, then the two must be comparable, and if comparable, perceivable. The internal constitution of the clock is examinable by removing the face. The distinction between primary and secondary qualities seems to be one only of measurable facility. The linguistic metaphor paints a picture that avoids some of these difficulties. Any event can become a sign. The movements on the face of the clock *suggest* its internal movements, and both are examinable, but it is only one clock whether we examine it with microscopes, X-rays, or telescopes.

A good metaphor may suggest improvements in technique. First, the aims of technique are *prediction* and *application*. Neither is present in the connotation of "cause-effect." Both are defining features of language-signs. Their two essential functions are: (1) that they bring us into touch with things remote in time or space, (2) so that we may be able to do something about them. Accordingly, the laws of nature are "a grammar for the understanding of nature, or that series of effects in the visible world whereby we are able to *foresee* what will come to pass in the natural course of things; which enables us to *regulate our actions* for the benefit of life." Thus prediction waits upon control of nature, and both are at the heart of the language metaphor. Secondly, while the search for an infinity of ghosts in the machine is an extravagant pursuit that clutters up procedure, the investigation of noticeable signs is more economical. It is surely cheaper to lay a multitude of unnecessary ghosts (*S, 252; P, 31*).

Which metaphor offers more fruitful suggestions for making an adequate theory of mind? According to the mechanical model the eye is a camera that takes inverted pictures of the external world. In one version the mind contemplates these pictures from which it makes inferences to external objects. But Descartes rejected half of this version: He retained the inverted pictures but gave them no use; the movements accompanying them act directly on the soul. However, the consistent Cartesian view would be the complete application of the mechanical model to every subject matter, including the human soul as well as the souls of other animals. Thus in these three versions of the model, the mind is either a ghost in the machine or nothing but a machine. We have seen how the linguistic model was used to untie a knot in the whole optic theory. In similar fashion it unties a knot in the theory of the mind. Just as we do not contemplate our retinas, so we do not contemplate our minds; and just as we do not contemplate pictures on our retinas, so we do not contemplate images or ideas in our minds—that is, if this last phrase is interpreted physically. Again, as in the treatment of vision, we shall be able to dispense with occult qualities and hidden forces and bring everything into the light of day, as it were. Minds will be those things that listen to the signs of a language so that they can do something about the things they signify. And minds, signs, and the things signified will have all the privacy and all the publicity that words and meanings have—privacy because no one interprets a noise or a color in quite the same way as I do, and publicity because the visual square and the visual kangaroo are given roughly the same meanings in Australia as they are in America.

Now it may be the case that the mind is a ghost in a machine, or better, it may be useful to make believe it is. The question becomes: Which is the more fruitful make-believe? The answer would involve testing along lines similar to those undertaken here for vision. But we can see at once difficulties in the "Two-World" view, such as that of finding our own ghost and the ghosts of other machines. Other difficulties ensue if we make believe that the mind is a machine. Eventually we should have to ascribe to it so many qualities ordinarily ascribed to persons and not to machines, such as learning, thinking, perceiving, interpreting, and reflecting, that it would be difficult to sustain the pretense that the mind is a machine.

From the consideration of these tests alone, it begins to seem as though the world may be illustrated just as well, if not better, by making believe that it is a universal language instead of a mighty machine. But there is a final test that will certainly carry weight with many. If our most holy religion is founded upon faith and not upon reason, as Hume claimed, then, other scientific things being equal, it would be more appropriate for us to choose that metaphor that fits with our religion. Let me extend the linguistic metaphor just as Newton extended the machine metaphor. Just as he believed that the "main business" of physics was to argue from the phenomena "till we come to the very first cause, which certainly is not mechanical," so I can argue from the signs, the things signified, and rules of grammar, to the conclusion that the events in nature are the language of the author of nature. In which case I have another *a posteriori* argument for the existence and nature of God. At least three courses would be open to me. First, I could lose awareness of my metaphor. I could treat this device as more than a metaphor, becoming so enthralled by the appealing picture it painted of a world in which the letters of divinity were written so largely, constantly suggesting the immediate presence of a benevolent deity who warned us beforehand of precipices, hurricanes, and approaching automobiles, that I mistook the metaphor for the face of literal truth. Secondly, I could treat the Myth of Language as Plato treated the Myth of the Earth Born, and perhaps as Descartes treated his Myth of the Machine: treating it as a modern allegory myself but offering it for literal consumption at a later time, knowing full well that the generation to whom it is first told cannot possibly believe it, but that the next may. If successful, I should attain to the full Wizardry of Oz, for the essence of full Wizardry consists in fooling others with our devices without being fooled by them ourselves.

But neither of these courses is acceptable; the first because the main theme of this book is that we should constantly try to be aware of the presence of metaphor, avoiding being victimized by our own as well as by others, even though my secondary theme is that the distinction between *bona fide* science, so-called, and scientific mythology is so tenuous that it cannot be sharply drawn; and the second because it is humbuggery. The third course is to be fully aware of the presence of the disguise; aware that there are no proper sorts into which the facts must be allocated, but only better pictures or better metaphors; and also aware that even to adopt a metaphor as metaphor is to alter one's attitude to the facts; and then to treat the language metaphor as a myth, "a myth not to be taken literally, but to be dwelt on till the charm of it touches one deeply—so deeply that when the 'initiated' say 'it is not true,' one is able to answer by acting *as if* it were true."[10]

10. J. A. Stewart, *The Myths of Plato*, p. 114.

Appendix

Models, Metaphors, and Formal Interpretations

1. Introduction

In recent years, the formal theory of models has come to be widely recognized as a powerful deductive discipline which has numerous applications to mathematics, philosophy, and the theoretical sciences. Many topics in abstract algebra have generalized counterparts in the theory of models, the most sophisticated formulations of empirical theories have taken the form of models, and model-theoretic methods have become indispensable in specifying the semantical interpretation of formal systems. In view of this wide-spread acclaim of model-theory, it seems natural that readers of the preceding pages should wonder what connections there might be, if any, between a "model" in the formal sense (which we shall call an "interpretation") and a "model" as characterized in this book. We make it our task to show here that there are indeed important similarities between the two notions. More specifically, we hope to establish that the philosophical concepts of model and metaphor can be explicated in terms borrowed from formal model-theory.

2. Theories, Interpretations, and Truth under an Interpretation

For the convenience of readers who are not already acquainted with model-theory, we shall describe, quite briefly and informally, some of the basic notions employed in that theory. For precise definitions and elaborations, we refer to the extensive literature on the subject.[1]

To begin with, we restrict our attention to those English sentences which have symbolic counterparts among the formulas of a

---

1. Clear introductory sections can be found in Benson Mates, *Elementary Logic* (Oxford, 1965), especially Chapter 4. More advanced topics are treated in Abraham Robinson, *Introduction to Model-Theory and to the Metamathematics of Algebra* (Amsterdam, 1965).

standard first-order predicate calculus. By a *theory* we shall understand any set of such sentences. Thus, any set of postulates or basic laws is a theory. Indeed, any set of sentences adequate to the description of any subject-matter will count as a theory. By the *vocabulary of* a given theory, we mean the class of all names, predicates, and relation expressions which occur in the sentences of the theory. For the sake of a more convenient exposition, we shall assume that the vocabulary of every theory comprises at least one name.

Given any theory *T*, an *interpretation I appropriate to the vocabulary of T* will be uniquely determined by the following respective constituents:

(1) the *universe of discourse* of the interpretation *I:* any non-empty set. Intuitively, a universe of discourse comprises any items of which mention can be made, according to the interpretation *I*, by means of discourse availing itself of the given vocabulary;

(2) the *naming-function* of the interpretation *I:* a rule which specifies, for every name in the vocabulary of the theory, which item in the universe of discourse shall be named by that name. Thus, if "Tom" should be a name in the vocabulary, and if Tom is an item of discourse, the naming-function might assign the person Tom to the name "Tom";

(3) the *assignment to predicates* of the interpretation *I:* a rule which specifies, for every predicate in the vocabulary of the theory, a class of items in the universe of discourse to which the predicate shall be applicable. Thus, if the predicate expression "is blond" should be in the vocabulary and if the universe of discourse should comprise persons, the class of all blond persons in that universe might be assigned to the expression "is blond" under the interpretation *I;*

(4) the *assignment to relation expressions* of the interpretation *I:* a rule which assigns to every relation expression in the vocabulary of the theory an appropriate relation among items of discourse. Thus, if the expression "loves" should be in the vocabulary of the theory and if the items of discourse are persons, then the rule in question might assign to the word "loves" the relation of loving (that is, the set of all pairs of lovers) confined to the universe of discourse.

Given any theory T and any interpretation I appropriate to the vocabulary of T, one can (recursively) specify, for every sentence constructed from that vocabulary, conditions under which the sentence shall be *true under the interpretation I*. Informally, the clauses in the definition may read as follows:

(1) If N is a name and P is a predicate, then the sentence formed by concatenating N with P is true under the interpretation I just in case the item named by N (according to the naming-function of I) is a member of the class assigned to the predicate P (under the interpretation I). Thus, "Tom is blond" might be true under I if the item of discourse named by "Tom" is a member of the class of persons assigned to "is blond."

(2) If $N_1, \ldots, N_k$ are names and R is a k-place relation expression, then the sentence formed by appropriately concatenating these expressions is true under the interpretation I just in case the respective items named by $N_1, \ldots, N_k$ (according to the naming-function of I) enter the k-adic relation assigned to R (in the interpretation I). Thus, "Tom loves Mary" might be true under I if the items of discourse named by "Tom" and "Mary" enter the relation assigned to the 2-place relation expression "loves."

(3) If M and N are names, then sentences of the form "M = N" are true under the interpretation I just in case some item of discourse is named both by M and by N (according to the naming-function of I).

(4) A sentence of the form "it is not the case that P" is true under the interpretation I just in case the subsentence P is not true under I.

(5) A sentence of the form "P and Q" is true under the interpretation I just in case each of the subsentences P and Q are true under I.

(6) A sentence of the form "everything is such that it is so-and-so" is true under the interpretation I just in case for some name N in the same vocabulary, the sentence "N is so-and-so" is true under all those interpretations J which differ from the given interpretation I at most with respect to the items assigned by the naming-functions to the name N. Thus, if the name "Tom" is in the vocabulary, the sentence "everything

is such that it has mass" might be true under I if the sentence "Tom has mass" remains true no matter what item of discourse "Tom" might name (according to naming-functions which are allowed to vary from that of I).

Since every sentence under consideration is equivalent to one of the sentence-forms mentioned in items 1–6, the notion of truth under an interpretation has, in effect, been defined for all sentences.

Given a theory T, we say that I is an *interpretation of* T if I is an interpretation appropriate to the vocabulary of T and every sentence in T is true under I. For every consistent theory there is an interpretation of that theory, and every interpretation is the interpretation of a "maximal" and consistent theory (comprising all sentences which are true under that interpretation).

Given a theory T and a sentence S constructed from the vocabulary of T, we say that S *is a (logical) consequence of* T just in case for every interpretation I (appropriate to the vocabulary of T), if all sentences in T are true under I, then S is also true under I.

This informal outline of basic notions is sufficient for our purposes.

3. Isomorphism and Perfect Models

In the preceding work, models have been characterized as certain kinds of metaphors; and metaphors, in turn, have been conceived as descriptions of a given subject-matter in the terminology appropriate to another. Since we have defined theories in a sufficiently wide sense to encompass any description of any subject-matter, metaphorical descriptions of a given subject-matter can tentatively be regarded as theories pertaining to that subject-matter whose vocabulary is not the customary one, but is borrowed from theories treating of different topics. If models are to be metaphors, we can expect that the vocabulary of a model will generally differ from the vocabulary of the theory modeled. To mention an example elaborated in the main text: a language model of vision will fit that vocabulary which is customary in descriptions of a language rather than that which usually serves in describing vision. But if the vocabulary of the model differs from that of the theory modeled, what is there about the model and the theory which determines that both pertain to the same subject-matter? Since the idioms employed in the model and in the theory may be completely dissimilar, it seems natural to look for the required similarities among the interpretations of these idioms.

One of the strongest relations of similarity among interpretations is that called "isomorphism." Roughly speaking, two interpretations I and I', appropriate to the respective vocabularies V and V', are said to be *isomorphic* if one-to-one correlations can be established between the expressions in the vocabulary V and expressions of the same grammatical status in the vocabulary V' and also between the items in the universes of discourse of the interpretations I and I' in such a way that the naming-functions, the assignments to predicates, and the assignments to relation expressions of the two interpretations assign to corresponding expressions correlated things, sets of correlated things, or relations among correlated things. If two interpretations I and I', appropriate to the vocabularies V and V', are isomorphic, if S is any sentence constructed from the vocabulary V and S' is the sentence obtained from S by replacing expressions in V by their correlates in V', then S will be true under the interpretation I just in case S' is true under the interpretation I'.

Given any reasonable criteria concerning the sameness or difference of subject-matters, isomorphic interpretations make different vocabularies appropriate to the same subject-matter. It is tempting, therefore, to identify models of a theory with those interpretations which are isomorphic to the customary or intended interpretation of that theory. It is in this sense, for example, that the system of natural numbers can be regarded as a model of certain theories which treat of the repeated concatenation of symbols.

However, very few examples of models, and probably none of those mentioned in the main body of this book, can be so construed that they meet the stringent requirement of isomorphism. Only ideally perfect models will actually be isomorphic with the intended interpretation of the theories to be modeled. For it is surely an idealization to suppose that models of a given subject differ only idiomatically from the intended theories of that subject. And models could hardly be of service in discovering a theory if we must suppose that they are already structurally identical with the theory to be discovered. Still, it seems fruitful to begin an analysis of models by mentioning criteria which are satisfied by idealized or perfect models; and to inquire next, as we now shall, in what respects models may fall short of that ideal without ceasing to play their intended role.

4. Models Which Are Adequate to Part of a Theory

In order to explore the ways in which good models may be allowed to differ from the idealizations just described, it seems helpful to consider some examples.

Suppose that we are interested in explicating the notion of an hereditary property of human beings. In considering this notion, we might be reminded of the principle of mathematical induction: "if zero has any property and if, assuming that any number has it, its successor has it, then all numbers have that property." This principle could suggest that we use the system of natural numbers as a model for the subject of our interest. To begin with, we might correlate numerical expressions with their intended counterparts. Perhaps we regard the names "Adam" and "Eve" as counterparts of the numeral "zero," we associate the expression "is a successor of" with "is a child of," and we correlate the predicate "is a number" with the predicate "is a person." With respect to this dictionary, the principle of mathematical induction can be translated into the assertion "if Adam and Eve have a certain property and if, assuming that any persons have it, their children again have it, then all persons have that property." Properties which meet this description may well be ones which we regard as hereditary properties common to all men.

Although this numerical model seems reasonably adequate to its subject, its similarities with intended interpretations of the subject fall short of isomorphism in various respects. Some of these differences could easily be accommodated by suitably modifying and weakening the notion of isomorphic interpretations. Among them, we mention that the correlation between the children of a family and the successors of a number are not one-to-one, but many-to-one; that, in the absence of infinitely many generations of human beings, we only succeed in correlating generations with the numbers in some finite initial segment of the number sequence; and the fact that the numeral "zero" was made to correspond with two names, instead of one. In order to illustrate other and more serious divergencies between good models and isomorphic interpretations, we shift to another example.

It has previously been suggested that the properties of wolves could serve as a model of the properties of men. Now, no matter how good this model may be, we cannot and do not expect that every property of wolves corresponds to a property of men; nor, conversely, that every description true of men can be discovered by noting a correlated truth about wolves. The model may well be deemed a good one if we succeed in translating just a few predicates applying to wolves into a few of the predicates applicable to men.

These considerations might suggest that a good model differs from a perfect one essentially in just this respect: that not all truths of the model have counterparts in the theory modeled, but only those which can be expressed in partial vocabularies of the two subject-matters. Thus, the model of wolves may seem a good one if just part of the vocabulary needed in fully describing wolves can be translated into part of the vocabulary suitable in describing men, and if just those truths regarding wolves pass into truths about men which can be expressed by means of this partial vocabulary.

However, a mere limitation of the vocabularies is not enough. For, even after suitable restrictions have been imposed on the vocabularies, it can happen that a model is a good one, and yet certain truths of the model (expressed by means of that restricted vocabulary) are found to be false or irrelevant to the subject-matter modeled. To see this, consider again the numerical model of hereditary properties. It may interest us to learn that the numerical truth "no successor of a successor of a given number is a successor of that number" passes into the truth "no child of a child of a given person is a child of that person." For this reason, we may want to retain, in the partial vocabulary of our numerical model, the expressions "is a number" and "is a successor of." The numerical truth "some number fails to be the successor of any number" is expressed by the same vocabulary (for, recall that we do not regard logical words as part of vocabularies). Its counterpart would be "some human being fails to be a child of any human being." Yet, due to the difficulty of drawing a sharp line between human and pre-human ancestors, we may not want to be committed to this latter assertion, or we may even regard it as false. Hence, even among statements constructed from the same vocabulary, we shall want to transfer some, but not all, truths of the model to the subject modeled.

Due to these difficulties, the notion of an adequate model is bound to be relative in at least the following three respects: (1) with respect to the subject-matter modeled, (2) with respect to a limited vocabulary which we expect to translate, and (3) with respect to a certain class of statements regarding the subject-matter to be modeled, between whose truth or falsity one expects to decide by appealing to the model. But since mention of the vocabulary may be part of the manner in which we delimit the class of statements to which the model shall be adequate, the relativization (2) is superfluous. The intended notion can be defined as follows:

Suppose that T is a theory whose vocabulary is V and that I is an interpretation of T. Assume further that V' is a vocabulary of such a sort that the expressions in V can be correlated one-to-one with expressions of the same grammatical status in V'. Let K be any class of sentences constructed from the vocabulary V, and let K' comprise those sentences which are obtained from the ones in K by replacing expressions in V by their correlates in V'. Suppose also that I' is an interpretation appropriate to the vocabulary V'. Then I' may be called a *model of the theory T (under its interpretation I) which is adequate in deciding the truth of sentences in K* if and only if the following condition is satisfied: given any sentence S in K, S is true under the interpretation I just in case the correlate of S in K' is true under the interpretation I'.

In order to obtain a partial illustration of the manner in which we conceive of models, let us consider the language model of vision which has been described in the preceding text. We seek to formulate a certain theory T concerning the relation between visual data and their objects. The vocabulary V of this theory may comprise the expressions "is a visual datum," "is a visual part of," "is a visual datum of," "resembles," and others which we leave unspecified. In attempting to construct a language model of vision, we notice that certain expressions which are used in full descriptions of a language do not appear to have obvious counterparts in the description of vision. For example, in describing a language,

we may want to say that new words of the language can be invented, or that the grammatical rules are conventional, while
nothing analogous seems to be true of vision. Accordingly, we discard from the vocabulary of our proposed model the expressions
"invented," "conventional," and any others which may be important in describing a language, but seem irrelevant for our purposes.
In the remaining vocabulary V' of the language model, we may
find such idioms as "is an expression," "is a syntactical part of,"
"denotes," and "resembles." Next, we correlate expressions in the
vocabulary V of vision with grammatically similar expressions in
the vocabulary V' of the language model. Perhaps we let "is a
visual datum" correspond to "is an expression," we translate "is
a visual part of" by "is a syntactical part of," we correlate "is a
visual datum of" with "denotes," and we associate "resembles"
with itself. Next, we try to delimit a certain class K of sentences
regarding vision which are constructed from the vocabulary V.
The sentences in question should be relevant to the theory of vision we have in mind, and they should also be ones between whose
truth or falsity we expect to decide by appealing to the language
model. Among the sentences in the interesting class K we may
find, for example, the assertion "some visual data are not data of
anything," the statement "visual data resemble their objects," and
the sentence "the object (if any) of a compound visual datum is
determined by the objects of its visual parts." The given correlations between expressions determine a translation of the sentences
in K. Let the class of these translations be K'. In K' we may find
the translated sentences "some expressions fail to denote," "expressions resemble the objects they denote," and "the denotation (if any) of a compound expression is determined by the deno-
tations of its syntactical parts." The language model of vision will
be adequate in deciding the truth of the sentences in K if every
true sentence in K' regarding language passes into a translation in
K which is true of vision, and every false sentence regarding language translates into a false sentence concerning vision. Thus, in
so far as we have specified K, the language model seems to be
adequate with respect to K.

It is of course customary to speak simply of a *good* model, and to
suppress relativizations, in cases where the theory to be modeled,
its intended interpretation, and the class K of statements which

are of interest in that theory are all assumed to be well-known. And we may say that one model is *better* than another for the purpose of modeling a given theory if (but not only if) one model serves to decide a class K of sentences pertaining to that theory which is larger than (and not a consequence of) the class of statements decided by the other model. In judging one model better than another, one may also have in mind aesthetic criteria (how poetic the metaphor underlying the model is) or pragmatic ones (whether the model is likely to be suggestive or whether it lends itself to teaching). But we are paying attention here only to the cognitive content of models. Given that a model is adequate in deciding the truth of a certain class of statements, a *fallacy* in the use of models consists in inferring that the model is adequate in deciding a wider class of statements or, conversely, that its failure with respect to a larger class shows its inadequacy with respect to the smaller one.

5. Models and Metaphors

Roughly speaking, we have characterized a model of a given theory as an interpretation of another theory part of which can be correlated, in a manner which preserves truth-values, with part of the theory modeled. In this sense then, it can be said that to model a theory is to represent part of the facts described by that theory in the vocabulary peculiar to another theory. This formulation corresponds closely to the description given of metaphors in the main body of this book. However, models were there defined as sustained or extended metaphors. Given our present notion of a model, in what sense of the word "extended" can models be regarded as extended metaphors?

There is surely some difference, with respect to their cognitive content, between good metaphor, bad metaphor, and outright misuse of language. Now it seems to us that one metaphor is cognitively better than another if it lends itself better to modeling than the other. Thus, to liken the human mind to a computer is a better metaphor than to compare it with a book because in the former case, but not in the latter, it is rather easy to see how one would go about constructing a promising model of mental activities on the basis of the suggested similarities. If this is so, then the metaphorical merit of utterances, and the features which distinguish metaphor from non-sense should be analyzable in terms of the models to which they give rise: a good metaphor is one which can be extended to a good model. But a bad metaphor which is "sustained" in either the sense of "used frequently" or of "vigorously defended," or which is "extended" in the sense of "detailed," "its features made explicit" does not thereby turn into a better metaphor. Hence, these are not the senses of "extending" which serve in characterizing metaphorical merit or the relation between metaphors and models.

Let us define a *metaphor* as a statement which suggests (or, whose full analysis asserts) that the intended interpretation of one partial theory serves as a model for part of some other theory with a different vocabulary. Thus, the assertion "man is a wolf" seems metaphorical in as much as it suggests that the intended interpretation of some theory regarding wolves can serve as a model which is adequate to a partial theory regarding men. Models, it seems, are indeed extended metaphors if by "extending the metaphor" we mean "constructing the model whose existence is suggested by the metaphor."

Steps to be taken in extending a metaphor may include statements which serve to correlate portions of the vocabulary peculiar to the two theories, and ones which make explicit which portions of the theory one intends to model. Thus, in extending the metaphor "man is a wolf," one may draw attention to the properties expressed by such words as "ferocious," "running in packs," and "attacking the weak." One may also mention that no attention should be paid to other words describing wolves, such as "four-legged," or "furry," and one may discourage unintended correlations, such as that between "running in packs" and "long-distance running in teams." Furthermore, in extending the metaphor, one may make explicit one's intention to use only a portion of the theory of wolves to account for a portion of the theory of man; for example, one might mention that one intends to employ just the hunting habits of wolves in analogy with the warring conduct of nations.

Strictly speaking, the metaphor "man is a wolf" does not actually represent facts of human conduct as if they were facts regarding wolves, if by "representing" we mean "describing." At best, this short statement can be said to hint at such a representation. The representing itself takes place in the model suggested by the metaphor. Accordingly, we would like to distinguish between metaphors and what we shall call "metaphorical descriptions." Given a certain theory and a class K of sentences of that theory, a (partial) *metaphorical description* of the subject of that theory is any non-empty class of statements which are correlated with ones in K and true in a model of different vocabulary which is adequate in deciding the truth of the sentences in K. Thus, a metaphorical description of man by means of sentences which pertain to the wolf model might be a class comprising statements like "wolves are ferocious," "wolves run in packs," and "wolves attack the weak." In describing facts of human conduct in language primarily intended to apply to wolves, vocabulary which is primarily intended to apply to men need not be used at all. For this reason, a metaphorical description of man, using wolf terminology, need not mention man at all. Mention of the connection between the two descriptions is not part of the metaphorical description itself, but rather part of the account specifying what subject the metaphorical description shall be a description of.

Assuming that we can compare the cognitive merit of models appropriate to a given subject-matter, we can say that one metaphor is *better* than another (with respect to that subject-matter) if the model suggested by the former is better than the model suggested by the latter.

6. Models as Tools of Discovery

In defining the notion of a model, we have used paraphrase of terminology which is customary in formal model-theory. Of course, use of this notion does not presuppose an acquaintance with model theory, just as the use of numbers does not presuppose competence in defining the abstract concept "number." Indeed, models all of whose formal features have been made completely explicit can probably only be found in highly theoretical disciplines. But conceptual devices whose outstanding features can be construed as corresponding to the ones of our formal notion are frequently used and seem to us powerful tools in the discovery of theories. We shall briefly describe the mechanism of their use by generalizing from a well-known example.

Suppose that we were interested in accounting for the motion of

gas molecules under varying temperatures and pressures. A preliminary survey of the subject-matter may suggest possible analogies with another subject-matter: the motion of billiard balls on a
table. In constructing the model at which the analogy darkly
hints, we proceed by tentatively "identifying" items of one subject
with items of the other. Thus, billiard balls might be "identified"
with molecules. This move can be construed as a proposed correlation between the predicates "is a billiard ball" and "is a mole-
cule." We notice quickly that certain features of billiard balls are
of no interest to us, and that certain properties of gases will not
have counterparts among the properties of billiard balls. Accordingly, certain idioms applying to billiard balls or to gas molecules
are discarded from the vocabularies. Thus, the expressions, "is
wooden" and "lie in a plane" are eliminated from descriptions of
billiard balls, and the predicate "is chemically indecomposable"
will be excluded from the vocabulary describing molecules. We
may or may not be in a position, at this stage of inquiry, to identify positively the expressions in the two vocabularies which we
intend to retain; and the theory regarding billiard balls which we
intend to use as a model may, but need not be, precisely given.
Perhaps we could only say, if pressed, that we have in mind a
theory of billiard balls which concerns their velocities and impact
under the assumptions of perfect elasticity, no friction, and ran-
dom motion, and that the relevant vocabulary of the model shall
be any minimal vocabulary adequate to describing that theory.
Still, if our model is to be useful, at least some analogies between
billiard balls and molecules must be sufficiently clear to suggest
correlation of idioms. We may observe, for example, that the net
force exerted by the billiard balls on the sides of the table increases with the velocity of the balls; and we may find this fact
analogous to the increase in pressure of a gas under an increase of
its temperature. For this reason, we may at least decide to correlate whatever idiom is used in describing acceleration with whatever terms are used in describing increase of temperature. Using
these rough guidelines, we can proceed to translate sentences
regarding one subject into sentences regarding the other, even
though the rule which governs the translation can perhaps not yet
be expressed very precisely.

Usually, the subject-matter of the model (or the aspects which interest us) are more familiar to us than that of the theory to be modeled. But this need not be the case. In modeling, we might learn just as much about the subject of the model as about the subject modeled. Thus, if Plato used the functions (or parts) of a state as a model of the functions of the soul, this sort-crossing may illuminate the subject of the state just as much as the subject of the soul.

The process of discovery proceeds by translating sentences which are known to be true or false of the one subject into sentences of the other subject whose truth or falsity is then conjectured and appropriately confirmed or disconfirmed. There can surely be no doubt that actual discoveries take place through use of this method. Indeed, efforts to translate from an appropriate model frequently raise questions regarding the subject modeled which otherwise might not even have occurred to us.

Some of the translations from model to subject may fail to preserve truth-values. If so, we restrict the class of statements with respect to which the model is deemed adequate. The value of the model, as a tool for discovering truths about the subject modeled, will depend on the theoretical importance of that class of statements which can be carried over without change of truth-value.

We have defined the notion of a model in such a fashion that the chief property which is preserved, under translations from the model to the subject modeled, is the truth-value of statements. We might also be interested in preserving other properties. For example, we might wish to preserve basic laws, so that laws of the model pass into laws of the subject modeled. Alternatively, we might be interested in probabilities, observational content, or other properties which might be preserved in passing from the model to the subject-matter. However, we rest content here with the treatment of one important notion of modeling.

7. Conclusions

In recent years, philosophers have tended to emphasize the virtues of precision rather than those of suggestiveness, and the importance of investigating constructed theories rather than methods which lead to their construction. This may be the reason why the notion of a metaphor (which smacks of poetic vagueness) and the subject of models (which seems to pertain to methodology) have not received the attention they deserve. As one of few exceptions to this generalization, the preceding book illuminates the role which models and metaphors have historically played in the development of great philosophical ideas. In this Appendix it has been our aim to show, furthermore, that the central concepts of that book lend themselves to a treatment which meets high standards of precision and exhibits close connections with fashionable interpretations of formal systems.

Index